AF712541

Robert Musil

On Mach's Theories

Philosophia Resources Library

Reprints, Translations and Commentaries
Relating to Austrian Intellectual History

Editors-in-chief:
J.C. Nyíri · Budapest (H)
Barry Smith · Manchester (GB)

Editoral Board:
Roderick M. Chisholm · Providence, RI (USA)
Rudolf Haller · Graz (A)
Allan Janik · Wellesley, MA (USA)
William M. Johnston · Amherst, MA (USA)
Christian Thiel · Erlangen

Philosophia Verlag · München · Wien

Robert Musil

On Mach's Theories

Introduction by G. H. von Wright

The Catholic University of America Press
Washington, D.C.

Philosophia Verlag · München · Wien

CIP-Kurztitelaufnahme der Deutschen Bibliothek

Musil, Robert:
On Mach's theories / Robert Musil. Mit e. Einf. von G. H. von Wright. [Transl. from the German by Kevin Mulligan]. – Washington, D.C. : Catholic University of America Press ; München ; Wien : Philosophia Verlag, 1982.
(Philosophia resources library)
Einheitssacht.: Beitrag zur Beurteilung der Lehren Machs ‹engl.›
ISBN 3-88405-044-3 (Philosophia-Verl.)
ISBN 0-8132-0586-7 (Cath. Univ. of America Press)

Library of Congress Cataloging in Publication Data

Musil, Robert, 1880–1942.
On Mach's theories.
(Philosophia resources library)
Translation of: Beitrag zur Beurteilung der Lehren Machs.
1. Science-Philosophy. 2. Knowledge, Theory of. 3. Mach, Ernst, 1838–1916.
I. Title. II. Series.
Q175.M982613 1982 501 82-74281
ISBN 0-8132-0586-7

Available in North and South America from The Catholic University of America Press, Washington, D.C.

Translated from the German by Kevin Mulligan
Originally published under the title
Beitrag zur Beurteilung der Lehren Machs
Copyright © 1980 by Rowohlt Verlag GmbH, Reinbek bei Hamburg
Copyright 1908 by Robert Musil

ISBN 3-88405-044-3
ISBN 0-8132-0586-7

© 1982 for the English translation by Philosophia Verlag GmbH, München
All rights reserved. No part of this book may be reproduced in any manner, by print, photoprint, microfilm, or any other means without written permission except in the case of quotations in the context of reviews.
Manufactured by Pera Druck, Hanns Haug KG, Gräfelfing
Printed in Germany 1982

Table of Contents

Musil and Mach

1

In 1903 Musil gave up his job as assistant at the Technische Hochschule in Stuttgart, moved to Berlin and began to study philosophy and psychology at the Friedrich Wilhelm University. Five years later he completed his studies and got his doctorate with a published dissertation on the philosophy of science of Ernst Mach[1].

Musil's main teacher in Berlin was Carl Stumpf, a former pupil of Brentano and Lotze. Stumpf was also a renowned psychologist and author of a two volume work on the sensation of musical sound, *Tonpsychologie.* Part of Musil's work in Berlin seems to have been done in Stumpf's institute for experimental psychology. His talent as engineer proved itself in the invention and construction of a machine (*Variationskreisel*) for rotating monocoloured discs so as to produce, to the eye, impressions of mixed colours. Musil's appreciation of Stumpf as a teacher is interestingly reflected in an entry in his diary of the mid-1930s when he was living in Vienna. An assistant of Schlick's, he writes[2], had been talking to him about the then current ideas of 'physicalism' in the Vienna Circle and their application to psychology. To this Musil remarks: "Wieviel genauer ist es doch in der Stumpfschule zugegangen. Diese nüchterne und wissenschaftliche Atmosphäre war doch ein Verdienst dieses Lehrers". It is not surprising that the philosophical psychology of the *Wiener Kreis* should have seemed to Musil artificial and barren. A contemporary school in psychology which impressed him more favourably and probably has also left an imprint on his writings as an author of fiction was *Gestalt*-psychology, associated chiefly with the names of Wertheimer and Köhler[3].

Musil, however, did not find work in experimental psychology congenial[4]. The subject matter of his dissertation is pure philosophy. We

[1] *Beitrag zur Beurteilung der Lehren Machs.* Inaugural-Dissertation zur Erlangung der Doktorwürde, genehmigt von der philosophischen Fakultät der Friedrich-Wilhelms-Universität zu Berlin. Berlin-Wilmersdorf: Dissertationsverlag Carl Arnold, 1908.

[2] Robert Musil, *Tagebücher, Aphorismen, Essays und Reden.* Herausgegeben von Adolf Frisé. Hamburg: Rowohlt Verlag, 1955, p. 451f.

[3] See *Tagebücher,* on a meeting in Vienna in 1911 with von Hornbostel and Wertheimer, and p. 291 and p. 631f. on Köhler.

[4] *Tagebücher,* p. 445: 'Wenig Freude am psychologischen Experiment'.

have no reason to think that the choice of topic was not Musil's own. We know from his diaries that he was already acquainted with and impressed by the work of Mach before he went to Berlin to study philosophy[5]. There was certainly an element of personal concern involved in his choice of a theme. Musil wanted to know whether Mach's claim was correct that the methods and results of exact natural science, when properly interpreted, would give decisive support to the positivistic philosophy which Mach was professing. Musil's answer to the question is No. Mach had not been able to defend his claim consistently. An examination of his arguments revealed inner contradictions[6]. Maybe the answer was a disappointment to Musil – and a contributory cause to his decision to give up continued academic work.

There were external complications too. Stumpf was not too pleased with the work of his student. His own opposition to Mach was deeper and stronger than Musil's. He was hesitant about letting the dissertation pass, and we are told that there were controversies[7] between the two men before Musil eventually, on 14 March 1908, was promoted to the doctorate.

For some years after his promotion, Musil continued to live in Berlin. He was offered a *Dozentur* in philosophy in the university of Graz in Austria, where Meinong was Professor. Musil, however, declined the offer. He moved to Vienna early in 1911 and took up employment in the Library of the *Technische Hochschule.*

After the dissertation, Musil did not publish anything strictly 'philosophical' of his own. There are a few reviews of philosophical and psychological books, and a long – and at the same time critical and understanding – essay from the year 1921 on Spengler's *Untergang des Abendlandes*[8]. It is hardly any longer possible to tell in detail to what extent Musil followed the changes in philosophy and psychology[9] in the decades between the two wars. I do not know that he participated in the activities of the *Verein Ernst Mach* or associated much with members of the *Wiener Kreis* when he was living in Vienna in the 1920's and 30's. But he is known to have been a frequent visitor to the house of the mathematician-philosopher Richard von Mises, when he again resided in

[5] Cf. *Tagebücher,* p. 37.

[6] *Dissertation,* p. 78.

[7] Karl Dinklage, 'Musils Herkunft und Lebensgeschichte' in *Robert Musil, Leben, Werk, Wirkung,* herausgegeben von Karl Dinklage, Zürich: Amalthea Verlag, 1960, p. 217. The information is from the psychologist J. von Allesch who knew Musil in Berlin. Details of these 'wissenschaftliche Auseinandersetzungen' are not known.

[8] 'Geist und Erfahrung, Anmerkungen für Leser, welche dem Untergang des Abendlandes entronnen sind', *Der neue Merkur,* March 1921.

[9] Cf. *Tagebücher,* p. 445: 'Geistiges Miterleben der Wendung in der Psychologie und Philosophie'.

Berlin in the years before Hitler came to power. Von Mises was a prominent member of the circle of empiricist philosophers in the German capital who closely cooperated with their Viennese colleagues. It can hardly be doubted that Musil was informed about what was going on in these circles. (Cf. above on his reaction to 'physicalism'.)

It would be particularly interesting to know whether Musil had read Wittgenstein and what his reaction was to the author of *Tractatus Logico-Philosophicus*. There seems to me to exist a great kinship between these two most remarkable men. Also, their life-curves show a striking resemblance. What Musil writes about feeling (Gefühl) and related psychological concepts in the unfinished parts of *Der Mann ohne Eigenschaften* is often astonishingly like the 'later' Wittgenstein's writings on these topics. I have seen one brief mention[10] that Musil had taken interest in the changes in Wittgenstein after the *Tractatus* – but I should regard it as practically excluded that he had seen or read any of the dictations or manuscripts by Wittgenstein which were in circulation in the 1930's. (Nor do I know that Wittgenstein had ever read Musil.)

2

The two philosophers who had most strongly impressed Musil were Nietzsche and Mach. If we had to mention a third, it would probably be Ralph Waldo Emerson. Musil's reading of Nietzsche goes back to 1898. His first acquaintance with Mach seems to have been in 1902 when he was living in Brünn in Moravia where a year earlier he had matriculated as engineer from the *Technische Hochschule*.

It would be tempting to see in Mach the source of inspiration for Musil's abortive venture into academic philosophy, and in Nietzsche the philosopher-poet who kindled the spark in Musil the novelist.

This judgement would not be entirely wrong. Certainly the influence of Nietzsche was much longer lasting and can be clearly seen also in the mature writings of Musil. Traces of Musil's reading of Mach may be discernible too – but at least to me they seem accidental and without deeper significance to the content of Musil's later thoughts.

When set in the proper perspective of the time, however, the combination Mach-Nietzsche is more significant than many a modern reader might suspect. The philosophy of Nietzsche can be associated with such attributes as 'subjectivist' and 'voluntarist', that of Mach with 'phenomenalist' and 'positivist'. Both pairs of attributes have an affinity

[10] By Ervin P. Hexner in 'Musils Interessenkreis' in *Robert Musil, Leben, Werk, Wirkung*, p. 143. It is not clear from this reference, however, whether Musil's interest concerned the changes in Wittgenstein's style of life or style of thinking.

with something which is sometimes also labelled 'idealism'. Mach and Nietzsche were further exponents of a *Zeitgeist* which can be characterized as post-Darwinian 'evolutionism'.

Nietzsche made no systematic effort to develop an epistemology or theory of knowledge. The scattered remarks on epistemological matters which are found in his writings show similarity with the 'phenomenalism' or 'sensualism' of Mach. The parallelism was noted in a work of the time, *viz.* Hans Kleinpeter's *Phänomenalismus*[11]. Kleinpeter also wrote studies on Mach's philosophy of science[12]. He is, incidentally, one of the very few authors, beside Mach himself, to whom Musil refers in his dissertation.

One sometimes talks of a Hume-Mach tradition in epistemology – represented also by Bertrand Russell in some of his writings, and later by the logical positivists. As far as theory of knowledge is concerned, Nietzsche too belongs in this tradition.

Round the turn of the century philosophy witnessed a reaction against the positivist epistemology in the spirit of Hume and Mach, as well as against various forms of 'idealism'. In the German-speaking world this reaction can be said to stem from the philosophy of psychology professed by Brentano. Meinong in Austria, Husserl and Stumpf in Germany, and the Pole Twardowski were outstanding pupils of this remarkable teacher. In the English-speaking world Moore and the early Russell represented a similar trend. The first part of Husserl's *Logische Untersuchungen* had appeared in 1900. Moore's celebrated 'Refutation of Idealism' was published in 1903. In neither is Mach directly a target of attack but the kind of sensualist epistemology which he represents certainly is. A most violent attack on Mach's 'idealism' was Lenin's *Materializm i empiriokritizizm* published the year after Musil's dissertation. In the philosophy of physics Boltzmann, Mach's colleague in Vienna, defended a 'realist' position in conscious opposition to Mach. Of the two great innovators in physics in the early years of the century, Planck followed Boltzmann, whereas the young Einstein was more a follower of Mach.

It is in the setting of these philosophical issues of the time: 'realism' *versus* 'idealism', 'phenomenology' *versus* 'phenomenalism' that one also has to place Musil against Mach. Like Boltzmann and Planck, Musil can be said to defend a realist position in the philosophy of physics. He criticizes, in particular, the 'fictionalist' and 'subjectivist' aspects of Mach's thinking. A crucial issue concerns the status of natural laws and whether there is a *physical* necessity ('in nature') or only a *logical* necessity ('in thought'). Musil argues against Mach in favour of the

[11] *Der Phänomenalismus, eine naturwissenschaftliche Weltanschauung.* Leipzig: Barth, 1913.

[12] *Die Erkenntnistheorie der Naturforschung der Gegenwart.* Leipzig: Barth, 1905. (Dedicated to Ernst Mach.)

notion of natural necessity. But, as we shall see, his argument is not convincing and contains an obvious *non sequitur*.

Basic questions in the general theory of knowledge are, on the whole, set aside in the dissertation. There are some very trenchant critical remarks on Mach's phenomenalism and on his efforts to overcome the mind-body dualism. But there is hardly a trace of defence of the act-object analysis of states of consciousness which is so central to Brentano and his pupils or, for that matter, to Moore. This is in line with Musil's aim as set forth in the concluding paragraph of the Introduction to the book. His statement is worth quoting here in full:

> "The only aim of the present work is to get as exact a view as possible of the inner consistency of what Mach says. If one wanted to take into account the truth of Mach's results rather than the rigour of the arguments for his views, a much more broadly based work of epistemology would be needed. The present work is intended only as a contribution to such a broader work. It avoids, as far as possible, taking up positions which would require justification by reference to any personal opinions and limits itself to the attempt to demonstrate, by way of immanent critique, that Mach's account contains, besides numerous positive features, so many contradictions or at least obscurities, that it is impossible to accord it any decisive significance."

One gets from these lines the impression that Musil was anxious to stress his unwillingness to commit himself to any *alternative* to Mach's philosophy. Considering this and also the fact that Musil's 'immanent criticism' of Mach is not always very convincing, one can well understand the reserved attitude of Stumpf to the dissertation. The merits of the work, it seems to me, lie in the concise and lucid presentation rather than in the criticism or attempted refutation of Mach's philosophy of science.

3

After having stated in the Introduction the aim of his investigation and summarized some of the main tenets of Mach's philosophy, Musil proceeds to examine Mach's 'biological' view of science as a process for acquiring and systematizing knowledge. This scrutiny is undertaken in the second chapter of the book. It leads to an important distinction (p. 24) between what Musil calls an 'indifferent' and a 'sceptical' interpretation of Mach's standpoint. On the first interpretation, roughly speaking, Mach's emphasis on economy, idealization, and search for invariance and permanence is only a description of the way science progresses and scientific knowledge accumulates. On the second, Mach's position is also thought to warrant far-reaching epistemological and ontological conclusions of a 'sceptical' nature about the foundation of knowledge and the criteria of truth in science. Under the 'indifference'-interpretation one can, on the whole, agree with the account Mach gives. An important aspect of the *Werdegang* of the exact sciences is thereby described in biological and psychological terms[13]. The 'sceptical' interpretation, however, Musil is inclined to reject: in no case does it follow logically from Mach's 'denkökonomische Betrachtungsweise'.

Mach himself is not very clear about his own pretensions. But that he, by and large, saw his position as a sensualist (phenomenalist, positivist) philosophy of knowledge with the 'sceptical' implications traditionally associated with such a position is all too obvious from many of his utterances. It is of some interest in the context to note Musil's reference to Kieinpeter (p. 26), who not only gave to Mach's view the 'sceptical' interpretation which Musil criticizes but also interpreted Nietzsche in a similar vein (above p. 10).

In the third chapter Musil gives an account of Mach's criticism of the 'mechanistic' world-picture of classical physics and of some of its key concepts – mass, energy, inertia, space, time, movement, temperature, *etc*. The account given of Mach's 'antimechanism' seems to me extremely good and Musil's own, on the whole positive evaluation of it (p. 36) agrees, I think, fairly well with the present standpoint in the philosophy of science. Of Mach's criticism of the key concepts Musil says, rightly I think, that it perhaps constitutes the most important part of Mach's achievement (p. 40).

The fourth chapter deals with Mach's criticism of causality. The idea that causality is obsolete in science and has to be replaced by the notion of functional dependence or relationship can be said to have been in the air at the time. To English readers it is probably best known from Bertrand Russell's famous paper 'On the Notion of Cause', published in the Proceedings of the Aristotelian Society for 1912-1913. The ideas of Mach and (at that time) also of Russell may be characterized as a consistent development of the criticism of causality by David Hume.

Mach's criticism is trenchant and still today of great interest. Musil concedes that from the point of view of the working scientist Mach's position contains much truth. But from the point of view of the epistemologist it leaves open crucial questions. In what way and in what sense do the functional relationships between the scientist's conceptual idealizations correspond to relationships between 'real' phenomena? In particular: Does Mach's criticism show that the idea of necessary

[13] Cf. Husserl's judgement on Mach in *Logische Untersuchungen,* Vol. I, Ch. ix. This comes very near to Musil's opinion. There is no mention of Husserl in the dissertation, however.

connections in nature must be banished from scientific thinking and regarded as an atavistic remainder from a more primitive stage in man's intellectual history?

The discussion of these questions is pursued in the fifth and concluding chapter of the dissertation. The gist of Musil's argument against Mach seems to be that Mach, by denying the existence of necessary connections in nature, is unable to account for the obvious fact – rightly emphasized by Musil – that "eine logische Verknüpfung nur dann einen Erkenntnisgrund abgeben kann, wenn sie durch eine sachliche Grundlage gerechtfertigt ist" (p. 57). I do not think that Musil's argument holds water, however. He thinks he can accuse Mach of inconsistency. Mach had agreed that science can be successful in its pursuit of laws only if there exist regularities in nature and that the predictability of phenomena on the basis of laws is proof of the uniformity of the world. Now Musil thought that if it is agreed that the equations or functional relationships which are the laws of nature have to correspond to regularities among the phenomena, then there must exist necessary connections in nature. At first (p. 67) he does not say this expressly. He says that "solange die Gleichungen tatsächliche gesetzliche Beziehungen ausdrücken --- weisen sie auf reale notwendige Verknüpfungen". This, presumably, only means that there is an 'Anschein von Notwendigkeit' in nature (p. 67). But later he goes a step further and says (p. 79) that Mach, by postulating lawlike connections between natural phenomena, is thereby also postulating necessary connections in nature. Musil is here identifying 'lawlike connection' with 'necessary connection'[14]. Before he had only said that the first 'hinted at' the second. For the step from this to an identification of the two he produced no argument. Yet the question whether the notion of natural law involves the notion of natural necessitation is the very question at stake in the discussion. Mach denied this involvement. Musil simply assumes it. But thereby he also begs the question – and his conclusion against Mach is a *non sequitur*.

Before his final return to the question of law and necessity, Musil had made a digression (pp. 70–75) into a related, yet clearly distinct topic, viz. Mach's sensualism (phenomenalism) and Theory of Elements. Some of Musil's observations in this context are in my opinion very well taken. Mach thought that the laws of nature ultimately describe relations between constituents of reality which he calls 'Elements'. What these 'Elements' are is, however, not made very clear. As examples, Mach mentions colours, tastes, tones, odours, (sensed) temperatures, etc. He calls them 'sensations' – but he also insists upon their character as a

[14] *Dissertation,* p. 79: "feste, gesetzliche, das sind aber notwendige Beziehungen in der Natur".

'neutral stuff' out of which both the mental (psychical) and the material (physical) aspect of reality may be constituted. (The position is also known as 'neutral monism'.)

Musil acutely observes (p. 71) that the 'elements' which are related to each other through the equations of physics are not sensory but conceptual units. Even if the 'raw material' of concept formation has to be given in sensory experience, the concepts themselves cannot be identified with 'bundles of sensations'. This is true of colours and tastes as well as of the more 'abstract', quantified concepts which occur in the functional relationships of natural laws.

Musil's criticism of the sensationalism of Mach stands somewhat apart from the rest of the content of the dissertation. In Mach's philosophy it occupies a central position. A few decades after Musil had criticized it in his dissertation, it experienced a revival, first in Russell's *Analysis of Mind* (1921) and later in the doctrines of some of the logical positivists. Its historical importance notwithstanding one has, however, the impression that it has now receded into obsolescence. This, however, is not true of Mach's philosophy of science in the more restricted sense, *i.e.* of what he has to say about the character and status of laws of nature, about the categories of causality and substance, and about the fundamental concepts of mechanics, optics, and the theory of heat. What makes Musil's dissertation interesting to a modern reader is that it concentrates on those aspects of Mach's thoughts which seem most challenging and fresh today and probably will in the long run be regarded as those of most lasting importance. Musil is, I think, far from always successful in his efforts to criticize Mach, but his exposition of Mach's thought is fair and lucid and the dissertation still makes good philosophical reading.

4

Of all the great writers of this century Musil is perhaps the one who is most deeply 'philosophical' in the true sense of this word. But what is fertile and original in his thinking is not found, not even in germ, in his dissertation on Mach. Musil's digression into philosophy after he gave up the career of an engineer for which he had been trained turned out to be a blind alley for his genius. It was on the other road which he entered at about the same time with the publication of *Törless* (1906) and the early plans for what eventually became *Der Mann ohne Eigenschaften* that his creative talent and genius found fulfilment. This is true, also, of the philosopher in him.

G. H. von Wright, 1982

I Introduction: Nature of our task

Today it is the word of the natural scientist which carries weight wherever epistemological or metaphysical questions come under the scrutiny of exact philosophy. The times are past when a picture of the world sprang full-blown from the philosopher's forehead. Making use of all the methods and results of exact inquiry, philosophy today is trying to reorganize its relation to the widespread regularities nature has been shown to exhibit, to reformulate its position on the old search for a correct interpretation of the concepts of substance and causality and its position on the relation between the physical and the psychical, and so on.

It is therefore a matter of some importance when the claim is put forward, and by a natural scientist, that in spite of its link with the natural sciences, this attempt to bring about a reorientation in philosophy – at least in the form it usually takes – contains nearly as many absurdities as it does philosophical assumptions; and when he substantiates this assertion with theses like the following, whose aim is to erect a wall between that view of the natural sciences on which the investigations of philosphers are usually based and the natural sciences as they really are:

1. Natural science only describes what has happened instead of explaining it. Natural laws in particular are no more than tabular descriptions of facts (or mathematical symbols which are the equivalents of such tables); theories in the natural sciences are merely the connexions we establish between such tables. Considered as explanations, theories are merely more comprehensive intelligibilities in place of less comprehensive ones. Neither an individual law nor a theory says more than would knowledge of the experiences they are based on, taken on their own.

2. Just as there are no explanations whatsoever, so – in particular – there are no causal explanations. If causal connexions did exist, they would, at best, only enable one to establish a certain concatenation of events; one would not be able to see what the reasons for the concatenation were. Exact science has, moreover, shown that there are no such things as causal connexions. In the earlier stages of their development, the natural sciences were obliged to seek such causal connexions but they are now completely free of the need to do so, except

for a few worthless and cumbersome remnants. Their real aim is to establish functional relations between facts, which do not present one fact as the cause of another, but merely make it possible to get from one fact to another by calculation, a relation which is completely reversible.

3. With the abolition of causality there disappears also an essential part of the meaning of the concepts of thing or substance; and the philosopher's hope of explaining the world of experience on the basis of concepts of substance and with the help of causal relations between things is in a sense split down the middle once the causal relation disappears.

But the destruction of the concepts of substance is also assured quite independently of all other considerations. For anything that could be said about substances could concern only their law-governed behaviour; but the laws supposed to express this behaviour have developed into merely functional descriptions expressing very much more general relations, from which the concepts of substance have been cancelled out, like intermediate substitutions which have ceased to correspond to anything.

4. In thus turning away from its traditional goals, science is by no means left without fundamental points of view of its own. Once science is seen to be no more than a means of mastering facts, made necessary by the struggle for existence, i. e. is seen within an evolutionary perspective, then the foregoing will become intelligible and all false presuppositions will disappear. The laws, concepts and theories of science appear as economical tools to help us adapt adequately to the practical demands arising out of our relation to our environment. To understand this goal is to understand everything there is to understand about the existence of science.

5. The great merit of this view is that the hopeless problem of the relation between the physical and the psychical turns out to be meaningless. The ideas of a world of bodies and of a mental world have their origin in certain instincts and are even of practical value at a primitive level of orientation in the world. But as scientific ideas they should not be valued any more highly than other ideas and their purpose should be sought only in their suitability for an economical orientation in the world. When they cease to be suitable, as is now the case, where they no longer correspond to the state of science and are a source of confusion, it is a methodological requirement that they be abandoned.

6. This is made possible by the knowledge that the functional relations on which the equations of natural science are based are in any case relations between sensations or – as Mach, in order to avoid any dualist overtones, calls them – elements.

Our ideas of bodies are based on some connexion between elements such as red, green, pressure, motion and it is merely another such connexion, more precise and more fruitful, beween elements which are however in principle the same as these, which finds expression in laws of nature.

Our scientific orientation in the external world, therefore, consists in nothing more than the search for equations between elements. This, then, is the cognitive ideal, one abstracted from the mature science of physics.

It follows that psychology too, to the extent that it aspires to definite scientific knowledge, can seek only to establish functional relations and that its concept of substance – the ego or soul – falls outside the scope of scientific treatment in just the same way as did the concept of physical substance. For Mach, sensations are the basic psychical elements and it is in terms of their mutual functional dependence that the scientific picture of the life of the mind is to be grasped. But, as we have seen, the elements of physical occurrences are also sensations and so physics and psychology turn out to have one and the same object.

The given, then, consists simply of elements in a great variety of different connexions; to do physics is to concentrate on some of these connexions; to do psychology is to concentrate on others, the difference between the two is merely a difference of approach; it as little introduces a gap between the physical and the psychical as does the fact that, in one set of circumstances, the reactions of a gas can be expressed by Boyle's Law and in another by the Mariotte-Gay-Lussac Law.

At this point, dualism ceases to exist as a problem for a critically rigorous methodology which has attained complete intellectual freedom and the difficulties of dualism turn out to be consequences of an unjustified refusal to abandon a primitive and irrelevant set of questions.

These, then, freely rendered, are the most characteristic principles to be found in the writings of the physicist Ernst Mach. It is already clear from this provisional survey that individual components of the theory are not new. Related ideas are to be found in older sensualist and positivist writings, particularly in the works of Condillac and Comte, and one is reminded of Hume's influence in the treatment of the problem of causality and substance. But what is most characteristic, indeed the key to everything else and at the same time, for historical reasons, the most disturbing aspect, is the heavy emphasis on the methodological standpoint – with its claim to a more rigorous unity – and the connected claim that Mach's position, far-reaching though the implications of his ideas are, is simply based on the sure ground of the natural science and on nothing else: "I only seek to adopt in physics a point of view that need not be changed the moment our attention turns to the domain of another science"[1], we read at one point.

Now both this appeal to the natural sciences and the restriction of knowledge to connexions between the 'phenomena' have been characteristic features of positivist philosophy since the days of Comte. But in part already while Comte's main work was in the process of appearing (1830–1842) and in part a little later, Faraday, R. Mayer, Joule, Rankine, Clausius, W. Thomson, Krönig, Grassmann, Redtenbacher among others were at work with a very different aim from that pursued by Comte, and even today the majority of physicists are by no means convinced positivists. (Proof of this is the opposition encountered by Mach's view in precisely these circles.) Thus in spite of occasional exceptions, positivism has remained more of a matter for philosophy, its central question has remained the epistemological discussion of the relation between subject and object and its main tenet the rejection of everything which is extra-mental. This has remained true of Avenarius, Laas, Schuppe, Rehmke and other like-minded inquirers of the present-day.

This situation determines our interest in Mach. For not only is he the representative of positivism with the widest audience, someone who at the same time has a genuine background in natural science (and as a scientist of importance) – working alone (for the search for contact with related ideas makes a late and infrequent appearance in his writings), he owes the stimulus for most of his ideas to developments in his own particular branch of science – but he is also the first person to take seriously the assertion that his (positivist) convictions were solely obtained by applying views which had proven themselves in the natural sciences and that they are no more than a result of the development of exact research. Mach, therefore, makes good in his own person what had hitherto been only a more or less empty claim and so makes it possible to find out whether positivism lives up to one of its most dazzling and appealing promises, the claim that it is merely the backwardness of philosophers which explains their failure to recognize the extent to which exact and fruitful science is already following in the tracks of positivist philosophy.

Our task will therefore be to find out for ourselves whether Mach does in fact arrive at his views as a logical consequence of a true or at least consistent view of natural science. At the same time this will shed light on the disturbing phenomenon of a natural scientist whose rejection of that contemporary philosophy which looks to the natural sciences for support is so total that he can say of his *Mechanics* that "its aim", in regard to the most important positions of such philosophy, "is to enlighten or, to put it even more clearly, is anti-metaphysical"[2]. And, "I have aimed at removing an old and stale philosophy from science"[3].

A circumstance which should be mentioned is the fact that Mach's epistemological and even the properly metaphysical comments are to be found in his writings not in a rigorous, methodical form but in the form of aphorisms. In addition, as in the works on mechanics and the theory of heat, they are merely dispersed here and there throughout the text. Thus in the first place it becomes necessary to single out and bring together all those ideas which belong together and I should like to emphasize that, in my view, once this is carried out the most important part of our real task is over. For then the connexions between the different arguments are visible and a very different view of their strengths and weaknesses emerges from that which is available when the individual ideas, isolated one from another, lead the somewhat irresponsible life of aphorisms.

The only aim of the present work is to get as exact a view as possible of the inner consistency of what Mach says. If one wanted to take into account the truth of Mach's results rather than the rigour of the arguments for his views, a much more broadly based work of epistemology would be needed. The present work is intended only as a contribution to such a broader work. It avoids, as far as possible, taking up positions which would require justification by reference to any personal opinions and limits itself to the attempt to demonstrate, by way of immanent critique, that Mach's account contains, besides numerous positive features, so many contradictions or at least obscurities, that it is impossible to accord it any decisive significance.

II The cognitive-psychological and economic approach

As a result of the aphoristic character of Mach's writings the assessment one makes of many of his claims will vary from context to context, so that sometimes the same statements have to be analysed from different points of view.

We shall begin with the point of view put forward by Mach and mentioned under (4) in the Introduction, according to which science should be considered as a phenomenon of economical adaptation, a point of view which is also closely connected with a cognitive-psychological mode of approach.

This approach to knowledge is important; for it lends a fascinating edge to everything Mach says from the very start. It is also related to familiar sceptical currents of recent times and is thus well fitted to attract and keep the reader's attention. It is, then, clearly important for us to determine whether all this is matched by the epistemological importance of this approach. And it can, I think, be easily seen that this is not the case. It is therefore quite possible to follow with great interest Mach's characteristically stimulating und plastic approach to the development and nature of natural science without feeling tempted to assume that this approach in any way prejudices decisions about the results of the epistemology and metaphysics to which Mach is opposed. In other words, the successes of a biological-psychological approach are without any further consequences for anything else.

Fundamental to this approach is the assumption – an assumption which we shall examine again in other contexts – that only knowledge of facts is of value for the physicist, that this is the main goal of physics, and that everything else is merely a roundabout intellectual way of obtaining and representing knowledge of this kind[1]. Thus we read: "If all individual facts about which we desire to gain knowledge were immediately accessible to us, no science would ever have arisen. Only because the memory of the individual is limited must the material be ordered[2]." This order is the goal of science[3]. The goal was originally a practical one and although specifically theoretical interests evolved subsequently, they too can be reduced to practical interests and considered merely as a roundabout way of satisfying these. "Every scientific interest may be viewed as a mediate biological interest", we are told[4].

The strong emphasis on the practical task of science turns out to be a direct consequence of the fact that Mach treats science, and indeed all human activity, from the point of view of self-preservation[5], in no way different from the activities of the lowest organisms. "All the processes of a living individual are reactions in the interest of self-preservation, and changes in ideas are merely part of changes in reactions[6]". But once the general notion of evolution is applied to science itself[7] it clearly follows that science has to be considered from the point of view of continuity and economy for these are integral parts of evolutionary theory; on the other hand, the very applicability of these points of view to thought itself provides retrospective justification for invoking the theory of evolution here.

As far as the first point is concerned, it is characteristic of the theory of evolution that it attempts to understand a creature's properties and reactions by reference to selective adaptation to the processes in its environment. it proves to be an empirical fact that this adaptation is economical and continuous: i. e. that once a property exists it cannot simply be pushed out of existence by other properties under new circumstances, but is rather subject to a gradual transformation which, in economiscal fashion, does not extend any further than is absolutely necessary[8]. As to the second point, it is fair to say that the whole of Mach's work is an attempt to demonstrate that scientific thought and its development do in fact illustrate these consequences of the theory of evolution.

The most important points can be summed up as follows:

1. The genesis of conscious life shows it to have the role of an instrument of economy; for if self-preservation requires the adaptation of a creature's reactions to external processes, then where the environment has reached a certain level of complexity the range of factual diversity will easily outstrip the number of biologically significant reactions, so that a whole group of loosely related facts, regardless of actual differences, will meet with one and the same reaction. Where this undifferentiated reaction suffices as an answer to the practical needs involved, the process as a whole is economic and economical. Our initial uses of concepts follow the same pattern. Facts "of like reaction" are grasped under one idea and associated with a single sign. Consciousness here belongs to the type of an imperfect physical apparatus which responds to the processes in the external world only to a limited extent and in certain directions[9].

2. Further functions of consciousness also show it to be an instrument of economy: for once in possession of a certain number of ideas it does not construct new ideas when faced with new facts but rather adapts to the new tasks those ideas already at its disposal. And this is done with the least possible expenditure of effort by retaining the original thoughts and modifying them only to the extent required in order to deal with the new demands. Mach describes this behaviour as corresponding to the principle of continuity or, to the principle of permanence and sufficient differentiation[10].

In this connexion Mach distinguishes:

a) the adaptation of thoughts to facts (which was mentioned above). In particular, he describes this as a picturing and modelling of facts in thoughts. For this is what makes possible an adequate adaptation to the environment, the relation required for self-preservation; "in order to put ourselves into a relation with our environment", says Mach, "we need some picture of the world"[11].

b) the adaptation of thoughts to one another: "Ideas gradually adapt to facts by picturing them with sufficient accuracy to meet biological needs"[12] but "of course the accuracy goes no further than is required by immediate interests and circumstances. Since these however, vary from case to case the results of adaptation do not match one another exactly. It is biological interest which goes on to bring about mutual correction of the resulting pictures so that deviations are adjusted in the best and most profitable way."[13] Mutual adaptation of thoughts is therefore the further task thought must solve if it is to attain full intellectual satisfaction[14]and "this requirement is satisfied by combining the principle of the permanence of ideas with that of their sufficient differentiation."

3. If human thought is generally of the same sort as the behaviour sketched above it must follow that the principles of economy and continuity are applicable to science. Mach nevertheless never tires of repeatedly citing examples to support precisely this important point. The following are typical.

a) Scientific thought itself is characterized by economy and continuity. Thus Newton imagines the planets to be projectiles thrown into the air and simply modifies the notion of constant gravity to obtain that of gravitation dependent on distance[15]. Fourier constructs a theory of heat conduction by modifying for his own purposes a theory of the vibrations of strings; a theory of diffusion is subsequently modelled on this, and so on[16]. Just as the notion of the rectilinear propagation of light was being entertained, refraction and diffraction were discovered. The original notion was retained and extended with the assumption of an index of refraction; but this in its turn had to be further specified with the assumption that a special index is necessary for each colour. Scarcely had it become known that light added to light increases its intensity when suddenly a case of total darkness was observed, etc. "Ultimately, however, we see everywhere in the overwhelming multifariousness of optical phenomena the fact of the spatial and temporal periodicity of light, with its velocity of propagation dependent on the medium and the period. This goal - to survey a given domain with the least expenditure of thought and to represent all the facts in it with some one single mental process – may be justly termed an economical goal[17]."

Scientific progress through the formation of hypotheses is characterized as a whole by continuity and economy. For hypotheses are initially drawn from the available stock of familiar experiences, their deductive consequences are then compared with the new fact and the hypothesis is modified to take into account the result of this comparison[18].

b) All the aids and devices on which natural science draws contribute to its economic character. This is particularly true of mathematics, the fruitfulness of which is due to "the great economy of its thought-operations"[19], but it is no less true of all heuristic methods. Their basic method is that of variation.[20]"The method of change or variation presents us with like cases of facts, containing components which are partly the same and partly different. It is only by comparing different cases of refracted light at changing angles of incidence that the common factor, the constancy of the refractive index, is disclosed. And only by comparing the refractions of light of different colours does the difference, the inequality of the indices of refraction arrest the attention. Comparison based on change leads the mind simultaneously to the highest abstractions and to the finest distinctions[21]." Because comparison[22] forms the core not only of all inductive procedure[23] but also of all experiments, all scientific methodology has continuity as its goal. For comparison aims at the recognition that the new consists of components, whether modified or not, of the old and is economic in just the same sense in which, as was noted above, the formation of hypotheses is economic.

c) Scientific results – concepts, laws, theories – are characterized by economy and continuity. The economic task of a law of nature is to eliminate the need to know mere individual facts[24]. A law dispenses with the need to pay attention to individual cases by bringing together typical cases with the help of one thought. When certain conditions hold, what is to be expected is limited and regulated by a law[25]; this law functions as a schema into which only the particular conditions have to be inserted. To recognize that a law is a special case of a more general law is to substitute a more inclusive schema for a less inclusive schema[26]. Where it is only necessary to pay attention to such a schema the memory is freed of an unnecessary burden; it possesses directions for deriving from the given schema the whole range of individual facts and more specific laws. What is true of laws holds too of concepts. A concept which has been brought to a high level of precision in science contains past work in a condensed and economic form[27], all relevant conceptual marks are incorporated into its definition and, since these are connected with one another by laws, the specification of one mark which has diagnostic significance can represent the whole complex. Thus one can say: "All physical laws and concepts are abbreviated directions, frequently containing subordinate directions, for the employment of economically ordered experiences, ready for use"[28], and in this economical ordering all "the puzzling power of science" is to be found[29].

At the same time these theoretical formations also correspond to the need for permanence. For it is in them – in constant laws and equations as well as in the fixed marks of concepts – that thought seeks to grasp those ideas which can be held on to permanently whatever individual changes may occur, ideas without which change would be incomprehensible and incoherent[30].

Comment: Insofar as it is possible to comment on these points without touching on special lines of thought which all require separate treatment later, the following needs to be said. From the point of view of epistemology, an approach like the above, with its emphasis on developmental, cognitive and psychological factors as well as on the economy of thought, may be either sceptical or indifferent. I shall call it indifferent where it is conceived of as an approach running alongside the properly epistemological investigation of the grounds and criteria of knowledge. I would call it sceptical the moment it is asserted that the latter investigation cannot for some reason be carried out and that only from the point of view of economy or by reference to biological and psychological reasons is it possible to decide what knowledge is. The germs of both views can be found in the above.

a) Indifference of the principles. This view follows already from the fact that it is possible to agree with these stimulating observations without either holding that the tasks of epistemology are thereby disposed of or, perhaps, that they are even remotely involved. The very statement of the problem suffices to bring out the difference. For if one wants to make such claims at this level of generality at all, then it has to be admitted that the psychological course of all correct and incorrect thought, judgement and prejudice, illustrates the principle of continuity wherever unusual circumstances do not interfere. But the question when a train of thought should be considered to be continuous and the question what external and internal circumstances lead to the continuous development of a train of thought, together with the question when the result of a train of thought – no matter whether its development is continuous or discontinuous, that is, economical or non-economical – should be accepted as true, these all express so many inner differences that the assumption that what we have here are two partially intersecting but mutually indifferent sets of questions must certainly be conceded to be possible.

But then the insight that natural laws serve to free memory of the burden of a number of individual facts and that the same is true of scientific concepts can have nothing to do with the question how such laws and concepts must be constructed if they are really to serve this purpose, or what sort of status or adequacy they acquire when the facts on which they are founded are taken into consideration. And similarly the fact that these laws are also connected one with another certainly has a practical value and one which it is economical to make use of, yet a number of questions remain unanswered. How, for example, do matters stand as far as the relation between the respective guarantees of such interrelated laws is concerned? What underlying real relations are involved when there exists a similarity between the laws obtaining in two otherwise separate groups of facts which allows them to be brought under common general equations (light, electricity and magnetism, for example)? Whether or how such questions are thought to be answerable, they are quite definitely not to be dealt with by pointing to the agreeableness of our being able to fit laws of nature into different theoretical contexts. The same holds of the concepts of thing, causality, force and so on. Either facts require the formation of such concepts or they conflict with the concepts. The question is always whether one or the other can be established. But independently of this question and prior to any answer agreement can easily be reached concerning the instinctive origin of these concepts and their economic value.

b) But there are also indications of a more radical, sceptical position. Consider for example the principle of permanence, according to which there are certain basic, instinctive assumptions which are simply given[31] and which are subsequently adapted to our knowledge of facts with a minimum of modification. We are told that the kinetic theory of heat and the conception of electricity as a substance owe the appearance of justification they possess and on which their existence is based to a mere historical accident. Even those theories which involve no metaphorical hypotheses, but are purely conceptual and quantitative, are coloured by the models which precede them just because they develop by refining already existing ideas[32]. Occasionally, it may be added, the direction taken by the development of a whole discipline might have been different had it not been for some relatively slight historical circumstance; quite different concepts and systems of concepts might have resulted[33], and so forth; so that, on this perspective, even the most exact sorts of concept formation appear to be "accidental and conventional".

In the face of such a demonstration – and I have no reason to doubt its validity – one might feel tempted to take a completely sceptical view of science and to connect the principles with this view. Obviously, if the products of science in the course of their development depend on individual, psychological influences and accidents and if even the factually given factor of adaptation can steer this development in quite different directions depending on the different particular constellations[34] (i. e. on those facts and aspects of facts available for comparison) then the suggestion might well be that science, as the outcome of such adaptation, is not something which could only exist in one form and not in another. Indeed, experience shows rather that adaptation allows its results a certain margin of free play without thereby being obliged to renounce its practical purpose; if everything which makes up our knowledge of nature is merely such a product of adaptation then it is no longer something definite and unambiguous but merely one historically understandable result among many other possible results. One might try to contrast this with the everyday opinion which demands truth of the results of the natural science, i.e. just that objective definiteness (in certain respects which have to be made precise) justified by objective necessity, which is here denied. On such a view, there would then be no solid, so to speak absolute truth but only truth which is relative in the sense that any opinion will count as true provided it fulfills its purpose of providing adequate orientation. In other words, there is no truth at all in the authentic sense but only a practical convention contributing to self-preservation.

In favour of this sceptical interpretation there is the fact that Mach says of a book by H. Kleinpeter, "The Epistemology of Contemporary Research in the Natural Sciences", that it is an account with which, in all essential respects, he is in agreement[35]. The general epistemological parts of this book are full of just those trains of thought we have outlined above[36]. Were one to argue that Mach's endorsement was perhaps overhasty, the fact nevertheless remains that his own writings contain a number of statements which tend to move in the same radical direction or are at least ambiguous[37]. Such a view, then, is by no means a free phantasy about possible interpretations of Mach's principles but has to be taken seriously.

Thus not only are there indications pointing to both of the views we have outlined but each enjoys a degree of textual support. Bearing our task in mind, we can proceed as follows. We want to know whether and how far these considerations provide a general foundation for the specific limitations Mach imposes on inductive knowledge. The interpretation of the principles according to which they are indifferent is of its very nature completely irrelevant. As to the sceptical interpretation, what interests us is not whether it is or might be Mach's opinion, but only the question whether it can serve as a foundation or only as a background for what comes later, in other words whether the general epistemological position is itself so firmly established that the sceptical attitude can or cannot simply be derived from it in particular cases.

There is absolutely no doubt about the answer to this question.

For, first, were these principles to be decisive by themselves they would have to be sharply formulated with just this end in view and their scope exactly defined. It would have to be shown in a systematic way that the principles suffice to guarantee at least one practically adequate inductive inference. And finally reasons would have to be brought forward which would exclude every other theory of induction aiming at a higher cognitive ideal[38]. No such general investigation, however, is to be found in Mach's writings. The textual evidence shows only that in certain cases Mach is inclined to make agressive epistemological use of his principles, as when, for example, he says that the question whether or not physical appearances are to be explained by reference to things and their relations is to be decided merely on the basis of the economies this would yield; although for certain problems it is, he thinks, more suitable not to do so. The texts do not show what the justification is for any given application of his principles. (Or, where attempts at justification are made they rely on special reasons which require separate examination in each particular case.) As long as the exclusive justification of the economical-biological approach is not demonstrated however, all appeals to it in the face of other methodological arguments remain irrelevant. Errors arise when conclusions are drawn on the assumption that it alone deserves consideration when all that has really been shown is that it too is relevant.

Secondly, it is also clear that the principles by themselves do not suffice to secure even that degree of scientific stability demanded by Mach; and that, on the other hand, when one takes this as a basis for interpreting the principles their supposedly sceptical significance disappears, leaving the textual support which forced us to mention this possibility in the first place as a number of isolated contradictions.

On the one hand, Mach says that every development of a scientific idea is economical provided it is continuous. But on the other hand he says that even where the same degree of continuity is present it is quite possible for very different results to be obtained. He therefore also calls for the best possible adjustment between the results of different adaptations[39]. This means, however, that continuity by itself and the economy it guarantees are no longer decisive for Mach, and where they might still appear to be decisive he would be caught in a contradiction. For, as he repeatedly makes clear, he does not wish his epistemology to provide any support for epistemological nihilism. "All points of view which are of value for the special sciences retain their validity", he says explicitly at one point[40], and wherever one looks in his scientific works one finds him concerned with just that unambiguous definiteness which can be established in so many different ways but never by mere continuity since, as he admits himself, continuity characterizes the development of both knowledge and error[41]. And although this continuity is admittedly not an unambiguously decisive requirement, Mach explicitly requires of science unambiguousness or univocity[42]and at one point refers to the latter as no more nor less than the aim of continuous adaptation[43].

But if one looks at this passage what one actually finds is a corresponding restriction in the principle of continuity as a result of the requirement of 'sufficient' differentiation. This is the genuinely decisive factor; an adaptation which is not sufficient or adequate is not an adaptation at all and so it seems that harmony amongst Mach's opinions has been restored. But what does the emphasis on 'sufficient' mean? It may mean, as we have seen, that the degree of adaptation is never more exact than is strictly necessary. But this means no more than that knowledge in the inductive sciences must in a sense be gained by working from the ground up, that what passes as truth today may be seen as an error tomorrow, that induction proceeds asymptotically, so to speak. But this is a generally admitted fact and has no specific connexion with the bio-genetic approach. The ordinary theory of induction is normally concerned to establish, what, at least at any given moment, is to count as sufficient. But here too there is no conflict with the consequences of Mach's principles of adaptation. For, according to Mach, adaptation is only adequate if it makes possible the reproduction and modelling of facts, i.e. if there is no contradiction between the intentions of thought and the facts to which they relate. It is just this agreement which is required by the normal view. Similarly, absolutely adequate adaptation – which should be considered as an ideal limit – could only be that adaptation which never leads to contradictions and which corresponds to all familiar and newly discovered facts in its domain. But this is just what is normally called knowledge or a truth. Only adaptation of this sort can be economic adaptation pure and simple, for any other sort must fail to deal with certain cases, must be insufficient, mislead thought and so become uneconomic. Furthermore, the criterion of economy is here of only secondary importance, for one must first know whether an assumption agrees with all experience, which just means that one must know whether it is true; only then can it unreservedly be said to be economical. Thus every trace of an account which would conflict with the usual theory of induction is abandoned.

What then remains of the second role attributed to economy, over and above mere continuity, has absolutely no specific character of its own which would distinguish it from ordinary views of the matter. The adaptation of thoughts to facts need not take place in only one way but will be carried out by different people in different ways. But "we will be able to compare these different scientific approaches with one another and to decide which is more economical than the others. Considerations of economy provide us with a valuable point of view enabling us to find our way around and organize our scientific activities"[44]. Thus, when contrasted with repeated application of the sine theorem, Gauss's dioptrics provides an example of economy[45]. Mach calls only the most complete and simplest description economic, that is when the smallest set of simple independent judgements has been found from which all the rest can be deduced as logical consequences[46]. For, as he puts it, "the mind feels relieved" not only "whenever the new and unknown is recognized as a combination of what is known, or the seemingly different is revealed as the same" but also when "the number of sufficient leading ideas is reduced and they are arranged according to the principles of permanence and sufficient differentiation"[47]so that the "economizing, harmonizing and organizing of thoughts are felt as a biological need far beyond the demand for logical consistency" and, on the other hand, "every avoidable incongruity or incompleteness, logical differentiation or superfluity of the describing thoughts means a loss and is uneconomic"[48]. Now insofar as it is not merely a confirmation of the previous train of thought discussed above, this is no more than a reference to the fact that, over and beyond questions of truth and falsity, there is an area to which belong such useful distinctions as those between simple and complicated, clear and obscure theoretical formations[49].

But then, as a result of the use he himself makes of them, the objective scope of the consequences of Mach's principles turns out to involve no more than ordinary views on the matter and their specific value is no longer that they ground knowledge but that they illustrate it *post festum*. And Mach himself says: "As a natural scientist I am accustomed to investigating individual questions ... and to move from these towards more general questions. I adhered to this custom in investigating the genesis of physical knowledge. I was obliged to proceed in this way because a general theory of theories was a task which was beyond me . . . I therefore concentrated on individual phenomena: the adaptation of thoughts to facts and to one another, thought economy, comparison, thought experiments, constancy and continuity of thought and so on. I found it both profitable and sobering to consider ordinary thought and all science as a biological and organic phenomenon with logical thought as an ideal limit case. I would not want to doubt for a minute that investigation can begin at either end. And, as this makes clear, I am perfectly capable of dinstinguishing between logical and psychological questions, a distinction I think everyone is capable of making who is interested in the light psychology can throw on logical processes. Someone who has once looked carefully at the logical analysis of what Newton says in my 'Mechanics' will find it difficult to reproach me with the attempt to run together blind, natural thought and logical thought. Even if we had the complete logical analysis of all sciences before us, the biological and psychological investigation of their genesis . . would still be needed; although this would not exclude submitting the latter in its turn to logical analysis[50]."

But with this Mach has said everything we wanted to hear on this topic and we can note once again that nothing has been demonstrated which would enable us to move on from the principles to what follows. Where such a step can nevertheless be made out in Mach's work there is, first, a lack of any objective justification and, secondly, he contradicts himself and the consequences of important components of his account.

III The opposition to mechanical physics. Criticisms of individual physical concepts.

"My exposition always starts from physical details and from there rises towards more general considerations", says Mach[1]; and since, as we have seen in the last chapter, these general reflections provide no clarification of the question whether and to what extent knowledge of nature is subject to more significant limitations than is normally assumed, we too shall follow the path which begins with the details. For even if no restrictions on the domain of knowledge, neither in breadth nor in depth, could be shown to follow from the general view of knowledge as a product of adaptation, the reverse nevertheless remains possible, that is, that particular restrictions placed on the natural sciences invest the more general points of view with a certain, perhaps high, degree of scepticism.

There are two trains of thought which we want to discuss first in this connexion: Mach's hostile attitude towards so called mechanical physics and his criticism of individual physical concepts. According to Mach, these theories, which are based on mechanical presuppositions, and these concepts lack all independent explanatory value; they exist merely as indifferent, economic representatives of the facts, as indicated above. In contrast both to the view which hopes to find in the hypotheses of mechanical physics the true course of events behind the phenomena[2] and to the (quite independent) attempts to grasp the true structure of this course of events by progressively refining the concepts derived from phenomena, this view of Mach's involves setting a limit to the cognitive ideal in the sense that what previously counted as an end is downgraded to the status of a mere means. Theory and conceptual systems are no longer the goals of enquiry but a means of mastering the facts; and with the demonstration that any function which goes beyond this is impossible and contradictory, the claim that science is merely a matter of an economic relation to facts – and that more than this is not possible – acquires a specific meaning.

Newton separated the result of analytic investigations of phenomena, that which can be derived with certainty from securely established facts, from the hypotheses which serve to explain phenomena but which are not themselves proven. It was in this sense that he considered gravitational acceleration – the inverse square law – and the agreement between the case where bodies fall to the earth and motion in the planetary system to be the result of analytic investigation; whereas the question how the action at a distance this involved could be more fully explained he took to be a hypothesis and a matter of mere speculation[3]. "But hitherto I have not been able to discover the cause of these properties of gravity from phenomena, and I frame no hypotheses; . . . And to us it is enough that gravity does really exist, and act according to the laws which we have explained, and abundantly serves to account for all the motions of the celestial bodies and of our sea[4]." But where he nevertheless invents hypotheses, as in the case of his theory of emission, he excuses his arbitrary assumptions on the grounds that his discoveries remain uninfluenced by theory and that he himself only adopts his theory as a useful explanatory device – not as an account of reality[5].

It is difficult to say whether this Newtonian *hypotheses non fingo* should be taken to signify no more than a methodological attempt to separate the secure goals of physico-analytic enquiry from the uncertain results of the philosophico-physical considerations erected on top of these – such a boundary would correspond to the level of knowledge at that time, but could gradually be changed to incorporate the "hypotheses" into what had already been proven. Or whether it is to be taken as dismissing hypotheses once and for all to a subordinate position[6]. Whatever the truth of the matter even Newton's famous contemporary, Huygens, had a quite different view of the importance of hypotheses and it was his view which was to prevail during most of the subsequent period. He wrote in his *Traité de lumière*[7]*:* "There can be no doubt that light consists of the motion of a certain substance. For if we examine its production, we find that here on earth it is principally fire and flame which engender it, both of which contain beyond doubt bodies which are in rapid movement, since they dissolve and destroy many other bodies more solid than they: while if we regard its effects, we see that when light is accumulated, say by concave mirrors, it has the property of combustion just as fire has, that is to say, it disunites the parts of bodies, which is assuredly a proof of motion, at least in the true philosophy, in which the causes of all natural effects are conceived as mechanical causes. Which in my judgement must be accomplished or all hope of ever understanding physics renounced."

In fact this goal of a 'true philosophy' remained for a long time that of all who sought to 'understand' natural phenomena[8]. Throughout the eighteenth century and nearly all of the nineteenth century the majority of the leading physicists were occupied in providing mental models of the processes behind appearances, processes which would explain these appearances. The main ideas employed in this attempt were force, movement and matter, the latter appearing in the different guises attributed to fluids, which were initially accepted and then rejected, and in the manifold forms corresponding to atomism, the continuity hypothesis, and as aether and so on.

It was indeed the great number of such theories, each different from the others, which disturbed the credibility of each individual theory. And the conflicts between these ideas reveal the remarkable fact that, where theories were given up, the downfall of an idea was only infrequently due to a demonstration of its impossibility; it was much more likely to be abandoned merely because alternatives were more suitable for the mathematical modelling which was desired at all costs. A defect of even those theories which remained was that, although they became more and more complicated in order to deal with ever increasing numbers of new facts, they were unable to provide an adequate explanation of these. Thus on the one hand hopes placed in these theories were increasingly disappointed and it became easier to see their most fundamental weaknesses – the obscurity of the concepts of force, matter and motion they employed as explanations. On the other hand, the retrospective recognition that the concern to produce mathematical models had been historically decisive made it seem reasonable to regard only the characteristic of economy – which they quite definitely had – as of any significance rather than any explanatory value. In this way, the confidential aspirations of an earlier age gave way to the very much cooler attitudes of the present day, typical of which is the attitude of Maxwell. One of the greatest promoters of the scientific modelling of mechanical hypotheses, he nevertheless wanted the intuitive representations they made use of to be considered as mere pictures. An even more pregnant formulation of the same point is due to Hertz, who expressly restricted the only remaining function of hypotheses to the requirement that they need be no more than pictures of the facts which, because the consequences of the pictures are pictures of the consequences of the facts, make possible a unified representation of the facts.

What Mach has to say on this subject is no exception to this general tendency and, historically, should be regarded as having contributed to it. His writings, however, contain little explicit opposition to hypotheses based on pictures. It was not necessary for him to do this because, as we shall see, he directly attacks the physical concepts on which these theories are based, and with the demolition of its foundations the collapse of a building follows immediately. This has to be borne in mind if a series of occasional attacks[9] is to be properly appreciated. Apart from these, his remarks can be grouped objectively as follows. Most important of all is, certainly, the repeated demonstration that one and the same group of facts can equally well be explained by different, even contradictory pictorial hypotheses, so that the question of the truth or falsity of such hypotheses cannot be decided[10]. But this means that they leave one completely free and so Mach adds to this demonstration the requirement that one should in fact only allow oneself to be guided by their usefulness (for a description of the phenomena) when choosing between hypotheses, particularly since he says, the favourite models of mechanical physics do not contribute in the slightest to our understanding of phenomena even where there are no other competing models[11].

Provided this is borne in mind, Mach's general remarks about the essence and nature of picture-hypotheses are immediately intelligible. Mach also employs the expression "indirect description" for these hypotheses, and one has such a description, he claims, when one says "fact A behaves not in just one but in many or all of its features like an already familiar fact B", whereby one appeals, "as it were, to a description which has already been formulated elsewhere, or one which has still to be precisely formulated"[12]. Thus one says that light behaves like a wave-motion or an electric vibration, a magnet as though it were charged with gravitating fluids and so on[13]. These are then essentially analogies, for "fact A is always replaced in thought by a different, simpler or more familiar fact B, which can represent A in thought in some respects but, for the very reason that it is different, cannot represent it in others"[14]. This is both the attraction and the danger of such hypotheses. They offer definite advantages as far as representation is concerned, in that they make possible a unified conception; and as far as the progress of induction is concerned, in their heuristic value.

"What a simplification it involves if we can say, the fact A now under consideration behaves in many or all of its parts like an already well-known fact B. Instead of a single feature of resemblance we are faced with a whole system of resemblances, a familiar physiognomy, by means of which the new fact is immediately transformed into an old acquaintance. Besides, it is in the power of the idea to offer us more than we actually see in the new fact initially, it can extend the fact and enrich it with features which we are first induced to seek from such suggestions and which are often actually found. It is this rapidity in extending knowledge that gives to theory a quantitative advantage over simple observation"[15].

On the other hand, there is also a danger in the assertion that two groups of facts are essentially identical when they are demonstrably only related to one another by analogy. Thus Mach says:

"Apart from the elements essential for representing the facts from which a hypothesis has been derived, the latter always or at least usually contains other elements that are not essential. For the hypothesis is framed on the basis of an analogy, an analogy whose points of identity and difference are incompletely known, since otherwise there would be no need for enquiry here. For example, the theory of light speaks of waves, whereas only periodicity is needed to understand it. These further, accessory elements, beyond what is necessary, are precisely the ones that are subject to change in the reciprocal action of thought and experience, until they are gradually eliminated in favour of necessary ones"[16].

"If now, as may readily happen, sufficient care is not exercised"[17] – if, in other words, too much faith is placed in the hypothesis and one is not prepared to abandon it in the face of contradictory facts – "then the most fruitful theory may under certain circumstances become a downright obstacle to inquiry"[17], something which has in fact happened in a series of historical cases[18].

If errors of this sort are avoided, there emerges in the course of historical development what Mach calls "the function of hypotheses" which "is, partly, to be reinforced and sharpened and, partly, to be destroyed"[19]. What this means is that when different hypotheses succeed one another all that remains is that in which they all agree, what is essential – the conceptual expression of the facts freed of all pictorial traces[20]. For anything more than this is in danger, as we have seen, of coming into conflict with new facts. The following examples may help illustrate the way this transformation proceeds. If two similar bodies stand in a relation of heat transfer then the following relation holds: the product of the mass and the gain or loss of temperature in each body is the same. This is what suggested to Black that heat could be seen as a fluid; as matters developed further this notion of heat as a stuff had to be all but completely abandoned; all that was valid in what remained was precisely the above mentioned relation[21].

Similarly, Carnot's view of his cyclic processes took as its starting point Black's notion of 'caloric' but the validity of his results remained independent of the latter[22]. Similarly, the idea that coloured lights are independent, invariable and constant components of white light survived in just the form discovered by Newton; but the view of them as a stuff, an idea added by Newton, was discarded[23]. And, as we have already seen, nothing remained of Huygens' wave theory of light but the fact that "the periodic properties of rays behave like geometrically summable segments in a two-dimensional space"[24]. In the same way, "the properties of the aether, or light-propagating space, which behaves in part like a fluid and in part like a rigid body, gradually found conceptual expression"[25].

Such a purely conceptual grasp of the facts, one no longer containing anything inessential, Mach calls a 'direct description'[26]. Once one has grasped the trend of the development in this direction, it is always advisable to replace the indirect description by a direct description as soon as it is possible to dispense with the heuristic services of the hypothesis[27].

"When a geometer wishes to understand the form of a curve, he first resolves it into small rectilinear elements. In doing this, however, he is fully aware that these elements are only provisional and arbitrary devices for comprehending part by part what he cannot comprehend as a whole. When the law of the curve is found he no longer thinks of the elements. Similarly, it would not become physical science to see in the changeable, economical tools it itself has forged ... realities behind the phenomena; .... as the intellect, by contact with its subject-matter, grows more disciplined, physical science gives up its jig-saw puzzle with pebbles and seeks out the boundaries and forms of the bed in which the living stream of the phenomena flows. The goal which it has set itself is the simplest and most economical abstract expression of the facts"[28].

Of course, the original pictures do leave certain traces, especially in the more important theoretical contexts. But this conceals no dangers, provided the fact that such theories are essentially based on analogies is emphasized sufficiently clearly. By ascertaining exactly in what respects a picture does and does not correspond conceptually[29] to a fact[30] there is no danger of taking it to be something real[31]. Used in this way analogy actually becomes an excellent means of mastering heterogenous fields of facts with the help of one unified conception and shows the lines along which a general phenomenology can be developed. Once made conceptually precise in the way required, analogy combines the essence of direct description with the convenience of pictoriality[32].

The most important aspect of these comments, as far as we are concerned, is that they display in a sharply focussed form one meaning of the principle of economy (or of continuity). As was already mentioned at the beginning of this chapter, Mach considers the original goal of mechanical physics to be as unattainable as it is without any purpose, so that only the economical suitability of its theoretical constructions for representing phenomena remains of value and deserves consideration. A presupposition which is here taken to be self-evident is that hypotheses must agree with the facts as far as their consequences are concerned; and this of course is decided not by reference to the point of view of economy but on the basis of the normal epistemological criteria. But beyond this it is economy which is decisive and nothing else and this involves a considerable narrowing of the cognitive ideal compared with older views.

Mach's position is nevertheless probably the prevailing one today and even if as far as mechanical physics is concerned the last word has perhaps still not been said on the subject, it involves so many questions that the scope of our task makes further comment impossible. Current opinion on the matter is so much in Mach's favour that we want simply to acknowledge this.

We pass now to the second part of our task, the discussion of Mach's criticism of individual physical concepts. First, a word about their importance. Mach speaks, as we have seen, about direct and indirect description. Occasionally direct description is opposed to indirect because it is conceptual[33]; this is misleading since both are of course conceptual. What is important about direct description is rather that it is the simplest conceptual description, i. e. a description no longer containing any inessential additions. But by inessential and accessory additions are meant those which are not guaranteed by experience and which may therefore also contradict experience. The sense of the distinction is therefore more exactly rendered by opposing what can be factually ascertained as the content of direct description to what is taken as a merely hypothetical basis or contributed by thought, on the basis of analogy, as an unproven extra completing the content of indirect description. The significance of the distinction lies entirely in the demonstrability which belongs to direct description. There is still no trace of any implication that facts, understood e. g. in the sense of what is real and is perceivable with the senses, should make up the content of direct description. On the contrary, the exact ascertainment of an analogy (think of the definition of analogy reproduced above!) is expressly referred to as being the object of direct description[34], as for example in the analogy between the behaviour of the planets and that of bodies swung round on a string with a certain tension[35]. By 'fact', then, is to be understood all secure knowledge and by direct description a proposition whose meaning does not in any way go beyond what is actually guaranteed.

But at this point a second train of thought intervenes and moves the emphasis even more in the direction of what can be sensed or intuited. For what is commonly taken to be factually ascertained has in many cases, according to Mach, no claim to this status. Even direct, conceptual description of facts, free of any hypothetical foundation is still not really what he calls the simple or simplest sort of conceptual description. This is only the case when the scientific concepts employed satisfy certain requirements. One need only think of the problem of causality which has been important ever since Hume, for whom only the spatio-temporal connection between certain events was factually demonstrable; the necessity of the connexion and of whatever else belongs to the concept of causality is, according to Hume, something merely added by thought. Mach extends this Humean position to other physical concepts – mass, energy, motion, temperature and quantity of heat amongst others. The everyday meanings of these concepts contain more than can be empirically demonstrated; the reasons for this are historical, psychological and economic. But if one wants to construct with these concepts a picture of the world which is correct, then this surplus, which is without any justification, and this picture must be kept apart. This separation is carried through by Mach for a number of examples and we want now to summarize his results.

Mach always has two points in mind in the following examples: first, the ascertainment of the factual basis of the concepts he is examining; second, the demonstration that all attempts to go beyond this basis and all arguments based on the results of such attempts lead to vagueness[36].

Thus in the case of the law of conservation of energy the factual basis is the observation that it is impossible that work be produced from nothing. In other words the most varied changes in physical states are brought about by mechanical work and where these can be completely reversed they yield the quantity of work required for their production[37]. But the fact that the energy can be converted into different forms asserts no more than just such an equivalence. To interpret this observation as if an indestructible something – energy – really remained conserved, and only the forms of its manifestations changed, is to interpret it as though a substance were involved. It is an interpretation which corresponds merely to the need in our thoughts for a stable view, an interpretation that we bring to the facts which, although they lend themselves to the interpretation, do not make it necessary[38]. Even as a possible interpretation its value cannot be rated very highly; the equivalence which was its factual basis depends entirely on the appropriate choice of the concepts of measurement for the magnitudes of the physical states which, together with work, are to be considered. Were certain magnitudes to be measured differently (and this seems to be a matter which is historically determined) not even this equivalence would be available and all support for the interpretation would be wanting[39].

Very much the same holds true in the case of the concept of quantity of heat; for certain vanishing thermal reactions equivalents occur[40] i. e. again, reactions which allow themselves to be considered as equivalents. But as in the opposed kinetic case there is a complete absence of any basis for an interpretation of this fact in terms of an underlying substance[41]. Each such interpretation is therefore to be kept apart from a simple reproduction of the facts. If clarity is to be obtained here it is necessary above all to put the concept of temperature on a firm footing, erroneous views of which have been the source of many sterile deliberations. The objection here[42] is to every sort of search for a 'natural' measure of temperature, for a 'real' temperature which is only incompletely expressed by the temperature we read off; the emphasis is on the fact that the measure of a body's thermal state – i. e. its behaviour in connexion with sensations of warmth – by any thermoscopic method is merely conventional, so that inferences from such measurements can only be drawn with this reservation in mind if they are not to lead to absurdities[43].

The elucidation of the concepts of space, time and motion is also extremely important. According to Mach, it is only as relations that they are guaranteed by expèrience. Whether a motion is uniform is a question that can only be answered with respect to another motion. Whether a motion is in itself uniform is therefore a senseless question[44]. Newton's attempt to distinguish between absolute and relative motion on the basis of the presence or absence of centrifugal forces[45] fails, since all he does is to distinguish between two groups of relative motion[46]. Since, therefore, all dynamic criteria drop out of consideration and since, from a purely kinematic point of view, absolute orientation is impossible, all that remains given for experience is relative motion[47].

But then this also knocks the bottom out of the concept of absolute space. It is admitted by Newton as far as absolute space is concerned that only relative positions are given in experience and the necessity of assuming an absolute space as the correlate of real motion is deduced merely from the dynamic differences of motion. But since Mach considers that these differences do not exist no support is to be found in experience for the concept of an absolute space; to insist on advancing this concept is to go beyond the boundaries of experience[48].

The same is true of time. Here too Newton distinguished between relative and absolute; relative time he regarded as the not quite exact measure (hour, day, year) of the absolute, true or mathematical time which appears in mathematical equations. Mach's objections to this is that the only factual, physical basis of the concept of time is the fact that the circumstances of some thing A change with those of some other thing B and depend on these. Thus, for example, the fact that the oscillations of a pendulum take place in time means no more than that its excursion depends on the position of the earth[49]. But it is utterly beyond our power to measure the changes of things by time alone[50]. Similarly there is nothing by reference to which an absolute time – independent of all change – could be measured; it is, therefore, a notion of no scientific value[51]. Absolute motion, absolute space and absolute time are mere mental constructs that have no demonstrable counterpart in experience.

To operate with concepts like these is to go beyond the boundaries of experience, which is illegitimate as well as meaningless: for there is nothing which can be said about such transempirical things[52].

Mach's attitude towards the concepts of mass and inertia is also important for what follows. To the Newtonian definition of mass as a quantity of matter, which is closely connected with the concept of substance, he opposes a demonstration that from such a notion it is impossible to obtain the meaning of the concept of mass: for this only the experience provided by certain experiments will suffice[53]. All that can be said is that mass m is assigned to a body if it imparts to another body, which is assumed to be a unity, m times the acceleration it receives under certain conditions[54]; and experience shows that this is the case for two bodies which turn out to have equal mass with respect to a third body and also with respect to each other and to other bodies[55]. "In such a concept of mass no theory is involved; the quantity of matter is unnecessary; all the concept contains is a fact which has been exactly fixed"[56]; "recognition of this fact is the furthest we shall go if we are not to fall into obscurity", says Mach[57]. And the entire significance of the law of inertia can be reduced to the same sorts of experience[58]. The law asserts no more than this, that it is accelerations which reciprocally determine bodies, under certain circumstances to be specified by experimental physics[59]; that in the absence of these circumstances there is no acceleration[60]; and that both propositions hold not only for the behaviour of terrestrial bodies relative to the earth but also for behaviour of the earth relative to distant celestial bodies[61]. It will be clear, after what has already been said, that acceleration here is always to be understood as relative acceleration[62].

These accounts make up what is perhaps the most important part of Mach's achievement, although the scope of the present work makes impossible anything more than the brief description given here. For in spite of their great interest and the fact that they take the reader straight to the heart of questions which are still hotly debated by specialists, our concern is only with their epistemological significance. And the position we must take on this is clearly and plainly indicated.

What has been demonstrated? That the central part of the content of certain physical concepts is gained from experience. But this is a triviality; as is the claim that the definition of such a concept "contains in a concentrated form a sum of experiences"[63] and that "all physical laws and concepts are abbreviated directions, which themselves frequently involve subordinate directions, for the employment of economically ordered experiences, ready for use"[64]. Mach has also demonstrated the failure of certain actual attempts to make a connexion between experience and physical concepts. These attempts sought to secure for certain physical concepts a meaning which, although it is supposed to be derived from experience (and may in this sense be said to count as the representative of this experience) does not merely express experience but also what is deduced from such experience (which need not, as is the case with absolute space, by any means belong to immediate sense experience)[65].

But what is the object of this demonstration? The first thing that needs to be said is that its function may be to connect up with the requirement that too hasty experimentation should be followed by attempts to form concepts in which for the time being one sticks as closely as possible to experience; and that one should go beyond the secure basis this provides as little as possible. Such a caution can never do any harm and may even in certain circumstances become a methodological requirement. We shall see later that this does in fact have points of contact with striking tendencies in modern physics. As a result of certain experiences, some aspects of which have already been mentioned in this chapter, physics today is attempting to protect its activities from the uncertainties of the metaphysics attaching to its results, by distinguishing as sharply as possible between that part of its concepts which is to be taken into account for purely physical reasons and everything else, and by concentrating only on the former. We shall have to come back to the limits of this tendency later, here it is enough to note that the only thing that could correspond to it in Mach would be the conclusion that there is a need for caution today for the reasons already mentioned; that explanations by hypotheses fail, that attempts to construct systems of concepts which tend to rise above the level of what is immediately experienced collapse; and that for the time being, therefore, it is open to us only to form our concepts in as naively empirical a fashion as possible. The fact that they represent experience economically is the only service we can safely expect of them today. – And with this, the issue raised by Mach would be disposed of[66].

But this is not the only tendency in the remarks we have quoted. There can be no doubt that Mach's criticism has a second meaning once one bears in mind the conclusions which, as we shall see, he draws from it, as well as the whole context. His criticism aims to show that it is absolutely impossible to deduce something from experience (and to form a physical concept corresponding to this in a meaningful way) which is not itself capable of being given immediately in sense experience. And this is the only possible interpretation of his remarks if, as has been mentioned, one is prepared to consider these in the context of the broad sweep of his thought, which from many sides issues in that sensualism according to which only sense phenomena are real and scientific concepts exist only to make possible orientation amongst these phenomena and can have no significance apart from this[67].

But this again is not what has been shown, however generously one may be prepared to determine the limits of what Mach has demonstrated. For difficulties and actual failures are by no means the same as impossibilities; after more experience has been gained the same attempts might be resumed and successfully completed. Thus in acoustics the tones are soon left behind and we proceed to deduce the vibrations causing the tones and wherever there is an absolutely satisfactory justification for going beyond immediate experience in this way this is perfectly legitimate, quite independently of whether what is deduced, such as the vibrations of sounding bodies, can or cannot then be given an illustration in intuition in some other respect[68]. The recommendation that all future attempts be declared meaningless merely because of the failure of some previous experiments has no scientific justification.

For it is important to be clear about the fact that what would then remain would be inadequate. Consider, for example, the definition of mass given on the basis of the facts alone; mass is there certainly something which is expressed only in relation to other bodies, but although these other bodies can change, the behaviour of the body under investigation remains such that the mass to be assigned to it always remains the same. But is this individual mass something which is peculiar to the body under investigation or to the bodies with which it is compared? I would not like to give any such simple answer to this question, which belongs to what falls under the concept of substance; but I wanted to pose the question for, whatever the answer, it is clear that the reaction comprehended by the sense of 'of mass X' is more closely connected with the body in which it is always found than with the bodies with which it is compared; the latter need be present only in the form of an arbitrary – not of any particular, individual – exemplar. But then it is at least something already to be found in the experiences to which Mach appeals that pushes toward forming the concept of a property. Space and time provide very similar cases. They seem, Mach says, to be quite special, because we can resort to different bodies for comparison and do not depend on any body in particular; and the mistake in this would be that one cannot make judgements about spatial and temporal behaviour [*Verhalten*] independently of comparisons with bodies – and just what would happen if one were to try this no-one knows[69]. But the reply to Mach's arguments here is as before. It is precisely the fact that one can speak of the same spatial or temporal behaviour even though one is making comparisons with different bodies (by which is meant that judgements about, for example, the time, could be made by reference to a clock, the earth's angle of rotation, a fall in temperature) which is evidence for the claim that such a behaviour is independent of the bodies we resort to for purposes of comparison[70]. (The fact that, where other bodies are completely absent, the original behaviour may no longer be presupposed is not a matter which belongs here but to the discussion of inductive methodology in general). The position is the same in other cases too.[71]

Everywhere, then, we find reasons impelling us to proceed further and what Mach demands would amount to ignoring these, which would be a highly unsatisfactory end-result. It might indeed be the case that a closer examination would make it necessary to adopt such a position, but no-one can be expected to hold it in the absence of conclusive reasons or if another alternative epistemological approach is available which avoids the main difficulties. But this position is precisely what we find in the case of Mach. The view that we find ourselves faced with a general, reciprocal dependence among phenomena which is nowhere rigid but more rigid here and there than elsewhere is already beginning to make itself felt here; likewise the hope that, even after taking account of this circumstance, it will nevertheless be possible to set out a securely established cognitive ideal; and the same is true of the prospect that, from the standpoint of this ideal, there will be absolutely no need to proceed further along the lines just indicated.

But this means that there has been a reversal of the entire situation: Mach's general views are not made necessary by the particular criticisms in the sense that the latter could somehow be deduced from these. Rather, the criticisms need the general positions if they are to lead to any definite result. Thus, at the end of this chapter, we can say that if arguments for the positions which are peculiar to Mach are forthcoming then whilst this may broaden the scope of those of his particular comments discussed here, on their own they have no independently decisive value as proofs.

IV The polemic against the concept of causality; its replacement by the concept of function.

The criticisms described in the last section were directed against particular theoretical structures; the criticism to be described now strikes at the very roots of the goals of scientific thought in general. Indeed this is the immediate result of having to abandon the hope of providing causal explanations for, at least so it is commonly thought, this is the purpose served by conceptual systems and theories. Helmholtz will serve as an example: the principle that every change in nature must have a sufficient cause forces us, he says , to infer the unknown causes of processes from their visible effects. Here, "the proximate causes which we attribute to natural phenomena may themselves be either variable or invariable; in the former case the same principle forces us to look for yet other causes of this variation etc. until finally we come to ultimate causes, which operate according to an immutable law and which therefore bring about the same effect at all times under the same external circumstances. The final goal of the natural sciences is, therefore, to discover the ultimate, invariable causes of the processes in nature"[1].

According to Mach, this is an unattainable and irrelevant goal. His reasons for making this claim are many and varied and their recapitulation here under their most important headings is intended to contribute to their elucidation.

1. Helmholtz's ideal of causal analysis seeks to identify ultimate causes which, where conditions remain the same, are followed by the same effect in an unequivocal fashion. This presupposes that such causes do in fact exist or, to use Fechner's terms, that in certain cases the same circumstances are everywhere and at all times really followed by the same result and where the one fails to occur so too will the other[2]. Mach objects to this that the same circumstances and the same results presupposed here can only exist in abstraction, in other words, only where other aspects of the facts are neglected, whereas in reality exact recurrence of like cases is not to be found. "In speaking of cause and effect"[3], he says, "we arbitrarily throw into relief those moments or features to whose connexion we have to attend in reproducing whichever facts we find important. There is neither cause nor effect in nature. Nature is there only once. The recurrence of like cases in which A is always connected

with B, i. e. of what is essential to the causal connexion, exists only in the abstraction which we perform for the purpose of reproducing the facts".

The real meaning of this objection penetrates to the core of the Machian theory of knowledge; for if it is justified, then not only is there no causal law 'in nature', but there is no law at all, for every law of nature seeks to express constant connexion. What is at issue here, in particular the role Mach accords to abstraction, is something we can discuss only later.

2. Part of the meaning of this objection lies in the claim that talk of cause and effect is based on inexact observation: for, it is argued, closer analysis invariably reveals the so-called cause to be only a complement of the whole complex of facts which determine the so-called effect. The complement in question will differ greatly depending on which components of the complex have been noticed or overlooked[4]. The rise in temperature of a body as a result of irradiation by the sun may serve as an example, the one is effect, the other cause. A more exact analysis shows that the influence of the environment and the intervening medium on the body's rise in temperature have to be taken into account; the irradiation by the sun is by no means the complete cause of the body's rise in temperature, but only one component thereof[5].

3. The connexion then is not simple but very complicated, indeed a whole variety of relations is involved. The relation between the sun and the body cannot be effectively isolated; the medium and surrounding bodies determine changes in the original body and are changed by it; and at the same time, there come into play similar relations in which they stand to an immense number of the other bodies[6]. The same is true of two heat conductors in contact or of two masses gravitating towards each other.[7] If we consider two gravitating masses or two heat conductors in contact as isolated phenomena, then the accelaration of the one is the cause of that of the other and vice versa, and likewise for the temperature changes of the conductors. But as soon as the ever present influence of other masses and bodies is taken into account not only does the relation cease to be reversible, it ceases also to be simple. Even in the simplest cases one obtains a system of simultaneous differential equations[8].

4. The relations which exact treatment of this sort brings to light do not possess the irreversibility characteristic of causal relation and do not express succession. Thus if one considers only the immediate relation between two masses or bodies in our examples, this turns out to be expressible by means of an equation, each element being obtained as a function of the other. But then cause and effect would be interchangeable

and so could not be characterized as cause and effect at all[9]. One says, indeed, that if a mass B comes into opposition to a mass A this is followed by a movement of A towards B; but this is imprecise and more exact examination shows that the masses A, B, C, D determine mutual accelerations in one another, accelerations which are therefore given as soon as the masses are posited[10]. Similarly, in the example of the sun's illumination of a body, were both to stand alone in an immediate relation to one another, the changes would be simultaneous and mutually determining; one could then regard a change in temperature of the body as the cause of the change in the sun's temperature[11]. And, again, when heat is transferred to a gas it seems that it may be considered as the cause of its tension but exact examination shows that both are variables of a single equation of state, changes in one variable lead to changes in the other and vice versa[12]. Mach summarizes this a follows: "Looking carefully at physical processes, we can, it seems, regard all direct dependences as reciprocal and simultaneous. For the ordinary concepts of cause and effect the opposite holds, because they are applied to wholly unanalysed cases of multiply mediate dependence": He goes on to illustrate this with the examples of a shot and the perception of a shining object. Between the explosion and the impact of the shell and between the shining object and the sensation of light there are intervening links, chains of mediate dependence. "The target that is hit does not restore the work done by the powder; the sensitive retina does not restore the light; both are merely links in the chain of dependences, which continue differently from the way they began. The target may yield flying fragments, the perceiving person may grasp for the shining object. The process as a whole need by no means be instantaneous and reversible just because it is based on a multiple chain of simultaneous and reversible dependences[13]."

Disregarding for the moment the full significance of the first objection above, the remaining arguments can be summed up as: the replacement of an approach in terms of causes by a functional approach:

"In the more highly developed natural sciences the concepts of cause and effect are constantly becoming rarer and more restricted in their use. There is a good reason for this: these concepts describe a state of affairs in what is at best a rather provisional and imperfect fashion because they are insufficiently precise...As soon as we can characterise the elements of events by means of measurable quantities, as is possible immediately for space and time and by detours for elements of sense-perception[14], the mutual dependence of elements is much more completely and precisely represented by the concept of function than by those of cause and effect[15]. This holds not only when more than two elements are in a relation of immediate dependence[16] but also and more importantly when the elements are in mediate dependence through several chains of elements. Physics with its equations makes this clearer than words can."[17]

In these words of Mach's the result of the objections raised finds expression: relations which are incompletely analysed are causal relations, relations which are completely analysed are functional relations.

If one goes on to ask what functional relations really are, the answer, as we have seen, is that they are relations expressing the reciprocal, quantitative dependence of the measurable components of phenomena; and we are referred to the equations of physics by way of elucidation. If now we consider one of these, for example that which holds between the pressure and volume of an ideal gas at a constant temperature, we see that in fact it contains no mention of succession and so no mention of causality. Instead of saying: fact B follows fact A and follows from fact A, the equation allows us only to calculate fact B given fact A. In other words, from the functional relation and the metrical characteristic of one follows the metrical characteristic of the other and vice versa, for in general B is just as likely to be a possible premiss for calculating A. It is in this sense, then, as we have seen, that functional relations are reciprocal, simultaneous and express no more than the dependence "of the conceptual elements of a fact in purely logical fashion" "just as they do for the mathematician, for example the geometer[18].

Without going into Mach's position on causality we can admit without further ado the comprehensive validity enjoyed by the concept of function on which Mach places so much emphasis. It is in fact a feature of every physical equation. It might be objected to the example brought forward, i. e. of Boyle's Law, that it expresses no causal connexion whatsoever but what, even on the ordinary view, is a simultaneous connexion, namely coexistence. But even laws such as Richmann's rule of mixtures, Galileo's law of gravity, or Kepler's law of refraction, and equations with such a specifically causal interpretation as those established by Newton to explain planetary motion[19], can all be taken functionally. Moreover the increasing prominence enjoyed in physics by representation in terms of differential equations seems also more and more to be forcing us to adopt the functional perspective. For whatever else one thinks about it, its great abstractness and its indifference – this is not meant in any adverse sense – to metaphysical problems cannot be overlooked. If in earlier accounts a central position was held by, for example laws of force, which when specified via the individual constants of the bodies involved yielded the different individual effects and thus allowed an explanation of individual phenomena, this appeared to have a very direct relation to reality and immediately suggested various metaphysical consequences. Today this step is no longer by any means apparent. The old laws of force usually occur today as special cases of more general laws which, expressed as they normally are by systems of differential equations, touch on metaphysics to a much lesser extent. For quite transparently they hang together with reality at one point only, by virtue of the fact that their consequences agree with reality. Physicists as important as Kirchhoff and Hertz have refused to accord them any other significance than this; and in any case the temptation to go beyond this indirect connexion between the conceptual content of these theories and reality, and to introduce a direct connexion is much less here than in the case of older theories, for the former offer far less support for such a connexion than the latter. One need think only of the flow of that vector which is characterised as the product of a force with a material constant and which forms one of the most basic notions in the theory of electricity. All that remains, then, is, in the first place, their suitability for representing phenomena, which they connect in a comprehensive fashion with one another – but without any indication of succession – so that calculation of the phenomena alone is made possible and so that they too fall under the concept of functional connexion. And it is to this that Mach then appeals[20].

This state of affairs is of the greatest importance for Mach and for the comprehension of what he says. Indeed I would even claim that it occupies a central position in the structure of his ideas. For it is here that the point of view of economy finds its strongest support; Hertz had broken with the tradition of starting out from individual propositions (the laws of force, the laws of conservation and so on) some of which were axiomatic, some of which expressed basic experiences, and of deriving everything else from these. For he pointed out that a self-deception is involved if one believes that fundamental equations which are supported by a few basic experiments enjoy a more exalted status than that guaranteed by the correctness of particular experiential propositions deducible from them. He suggested that, once in possession of basic equations (and today these are the differential equations just mentioned), we should cease all derivation, simply accept them and see their justification merely in their ability to represent the facts. Together with its abstract conceptual content, which, as we saw above, has only an indirect connexion with reality, we have the core of the principle of economy. For if we were able to note above merely that the economic approach is one factor among others which come into play, that one task of science is "to replace or economize on experiences by reproducing and anticipating facts in thought[21]", we can now substitute 'the only' for this 'one'. Functional relations make possible the logical determination of "the mutual dependence of features of facts on one another[22]", the most general equations subsuming them are merely "abbreviated directions for the employment of economically ordered experiences[23]"; "reproduction in thought is the goal of physics; atoms, forces and laws, on the other hand are merely means which make this easier; their value extends as far as their usefulness[24]". In other words, any other role is incompatible with the subtle process of modelling as this is carried out today[25].

This situation also deepens the force of the critique of different concepts sketched in the last chapter (and so the economic point of view). The concepts of natural science acquire their content from experience and from the regularities given through experience. Although one speaks of mass, force and heat-states etc., "no more should be understood here than the total behaviour which previous experience shows can be expected. One gives this a name or connects a certain image with it, but this is only to represent familiar processes, no more. Nothing can be deduced or inferred from it which is not the product of experience[26]". This is what Mach says and we agreed with him that in an empirical science this is decisive as far as the disposition of its concepts is concerned. But we did not find that he had shown that this is all there is to it and that the significance of concepts is that they simply indicate the experiences they represent. On the contrary, we found this requirement untenable in the absence of further support. But such support is now beginning to emerge from these arguments.

Of his definition of mass, Mach says[27] that it is intended "to establish the interdependence of phenomena and to remove all methaphysical obscurity without accomplishing on this account less than any of the more usual traditional definitions" – and the same is true of his definitions of the other concepts. This interdependence of phenomena is a functional dependence. And, according to Mach, it shows no sign of that tendency to go beyond mere description of the phenomena which we brought forward as an alternative to Mach at the end of the last chapter. For if the equations express no more than a connexion which makes possible the calculation of certain features of phenomena from others, and if the definitions of the concepts are contained in the equations then it seems that their meaning is exhausted by the fact that they express or serve such a connexion between appearances.

This requires particular emphasis in the case of the concepts of force and thing, for it is on those purified properties of things which are forces that their causal effectiveness is based. Where these disappear or cease to be more than representations of functional relations which in addition contain more, and more that is accurately expressed, than do those incomplete accounts which belong to a primitive level of knowledge, then the formation of causal relations becomes completely pointless. And Mach of course does not hesitate to apply his views to these concepts too.

Already in the original (mechanical) sense, force is defined as a circumstance which determines acceleration[28]; it is a concept which has nothing to do with the unknown causes of processes in nature; it signifies nothing hidden, but rather an actual, measurable circumstance of motion, the product of the mass and the acceleration[29]. When a body exerts a force on another body this means no more than that when the two bodies come face to face the second body exhibits a certain acceleration with respect to the first; the concept of force therefore expresses no more than a certain constancy of connexion[30]. (And the other concepts of force are merely versions of the original one.)

Such constancy of connexion is at the heart of the concept of substance; we term substance what is unconditionally constant[31]. Now there is in fact no really unconditional constancy; all cases of constancy are constancy of connexion or relation.[32]When, for example, we are told that the same body which with chlorine forms cooking salt forms Glauber's salt with sulphuric acid this signifies constancy of connexion or the interdependence of certain reactions: a class of objects A yields reactions a, b, c. Further observation perhaps reveals reactions d, e, f. If, now, it turns out to be the case that a, b, c, on their own unambiguously characterize the object A and d, e, f likewise, then this establishes the connexion between the reactions a, b, c and d, e, f in the object A[33]. This "constant connexion between reactions expounded in the propositions of physics represents the highest degree of substantiality that inquiry has thus far been able to reveal[34]." "When an equation is satisfied, then there is involved therein a widened and generalized concept of substance. In general it matters little whether we regard the equations of physics as expressions concerning substances (laws or forces), for at all events they express functional dependences[35]."

This, then, is the refined concept of substance of the natural sciences; it is here that the ordinary concept of a thing receives its conceptual modelling with results which hold retrospectively for the ordinary concept too. Thus if no real constancy is expressed by the scientific concept of substance – i.e. if the constancy it expresses is merely that of a group of functional dependences, of reactions which occur "here and there", and not that of a spatio-temporally individuated unity[36] – then this will naturally be all the more the case for the vulgar concept of a thing. It too expresses only a relative constancy. There is no unchangeable thing in nature; a thing is an abstraction, a symbol for a relatively stable complex; although it is subject to change we abstract from this[37]. One element of the complex disappears, then another, one element appears in a changed form never in fact to recur in precisely the same form, and it is only because the transition from one to the other is so smooth and because the total of constant elements at any moment is greater than that of the variable elements that it is possible to believe that something would remain even if all elements were to disappear; only in this way can the idea arise of a substance distinct from its attributes, of a thing in itself[38].

Thus what we learnt in the last chapter holds true in this case too: "all physical laws and concepts are abbreviated directions, frequently containing subordinate directions, for the employment of economically ordered experiences, ready for use[39]"; and if it applies here in the sense of 'mere economy' then this is because of the general critical meaning, which has already been described, of this passage[40].

All this is further strengthened by the fact that even the equations on which the formation of concepts is based are only relatively complete, as analysis of their meaning shows: first of all, in the case of the material constants occurring in these:

"The equation pv/T=constant holds", says Mach, "for a gaseous body of invariable mass for which pressure, volume and temperature have the same values in all its parts and provided the conditions are distant enough from liquefaction. The limitation contained in the law of refraction $\sin\alpha / \sin\beta = n$ involves further restrictions: to a definite pair of homogenous substances, at a definite temperature and density or pressure as well as to the absence of internal differences of electric and magnetic potential. If we apply a physical law to a definite substance, this means that the law is valid for a space in which the known reactions of this substance are also found. These additional conditions are usually covered and concealed by the mere name of the substance. Physical laws that hold for empty space (vacuum, aether) always and only relate to definite values of the electric and magnetic constants, and so on. By applying a proposition to a given substance we introduce further determinations (or equations expressing conditions) just as when we say, or tacitly assume, of a geometrical theorem that it applies to a triangle, parallelogram or a rhombus[41]."

The same is true, secondly, of forces – in precisely the same sense. Mach adds to what we have already heard: "It is part of the general idea of the Galilean-Newtonian system of mechanics to conceive of all connexions as replaced by forces which determine motions required by the connexions; conversely, everything that appears as force may be conceived[42] to be due to a connexion . . . If we reflect that in both cases, whether forces or connexions be presupposed, the actual dependence of the motions of the masses on one another is given for every instantaneous conformation of the system by linear differential equations between the coordinates of the masses, then the existence of these equations may be considered the essential thing[43]."

The aim of contemporary physics then is to represent all phenomena as functions of other phenomena and certain spatial and temporal positions[44]. But as we have seen, space and time are themselves concepts for certain connexions between phenomena: the oscillations of a pendulum, for example, take place in time only if its excursion depends on the position of the earth[45] and so here the measurement of time amounts to measurement of angles or lengths of arcs[46]. If we imagine the natural course of different events represented by equations involving time, then time may be eliminated from these equations (for example, an excess of temperature may be determined by space traversed by the falling body); the phenomena then appear simply as dependent on one another[47]. It is therefore completely superfluous to emphasize time and space since temporal and spatial relations merely reduce to dependences between the phenomena[48].

Thus the equations of physics refer to a very general connexion. For to be a function of time now means to be dependent on certain spatial positions; and that all spatial positions are functions of time means that from the point of view of the cosmos all spatial positions depend on one another; but since spatial positions can only be recognised by reference to states we can also say that all states depend on one another[49]. In our ideas of time, then, the profoundest and most universal connexion of things finds expression[50]. The same is true of our ideas of space, for every motion of a body K is a motion towards other bodies A B C...[51] and even if one says that a body preserves unchanged its direction and velocity in space this contains a reference to the need to take into account the whole world[52].

Let us summarize: We have already conceded that the concept of function is the real vehicle of modern physics; that the basis of concepts must be sought in experience; that the equations describing these experiences are first and foremost functional equations; and that we cannot ignore the fact that force, thing and causality occupy only a very unimportant position in scientific expositions or, at least in their original form, vanish completely.

But what does this amount to? Such concepts cannot easily be avoided, the complications of an alternative mode of expression prevent this; Mach too makes use of them for 'everyday purposes[53]'. On the other hand, these concepts belong to the philosopher's sphere of influence as well as to that of the physical scientist. There is, therefore,quite apart from their unequal successes, a strong methodological reason for separating the shares the two disciplines have in these concepts. It is thus completely natural for the physical scientist to seek to secure his results from philosophical surprises and make his laws, forces and material constants etc. independent of any further philosophical discussion and justification. This is most naturally done by drawing a sharp boundary and, for example, saying that whatever this X turns out to be, for me, the physicist, it is no more than that which functions in such and such a way in my equations.

Such tendencies are not new. Newton had employed the word 'force', strictly limiting its sense to that of the unknown cause of familiar processes; his aim in doing so however was not to anticipate further discussion but only to fix results already obtained so that whatever foundations they were provided with in future their existence would remain secure. Fechner, too, writes that force in physics is no more than an auxiliary expression for the description of the laws of equilibrium and motion which hold whenever matter and matter come together; the physicist knows nothing of force other than his knowledge of laws, and it is only in terms of this knowledge that he can characterize it. And, more recently, there is Kirchhoff who, tired of the sterile dispute about force and matter, their relation to one another, their nature and the like, excluded these questions from mechanics (their native soil) by setting this science the task of providing the simplest and most unequivocal description of the motions of bodies and by using force as the term for certain algebraic expressions which recur repeatedly in the description of motion rather than for a metaphysical cause of motion. Hertz's account of mechanics belongs here too and, in part, the approach to physics in terms of energy, Maxwell's theory of electricity, and others.

But however much this seems to support Mach's case, and although he appeals not only to modern modes of exposition but even to direct statements of Kirchhoff and other physicists[54], there is nevertheless in every case a great difference between what Mach appeals to and his own expressed aims. For when one says, "As a physicist I can only concern myself with this topic in such and such a sense" this signifies merely a change in one's task, it is far from signifying a change in one's subject-matter; other interests are by no means excluded; the emphasis on the specifically physical point of view involves nothing like an *anti-metaphysical* tendency.

This brings us to what is specific and truly decisive in Mach's position, to what is peculiar to his position alone and not simply to modern physics. He says of the concepts under discussion not only that they vanish from the face of physics but that they drop out of consideration absolutely; the situation is such as to make them both impossible and superfluous, for the scientific world picture is complete and self-contained without them.

The reason for this is, we are told, to be found in functional equations.

But, we may ask, what contribution can really be expected from these? Their purpose is to make possible the calculation of attributes from other attributes; on Mach's interpretation this dependence appears to be merely logical and instead of a cause only the role of a cognitive ground remains. But this view of the matter is incomplete. For it is clear that even the connexion expressed in a functional equation corresponds to a real dependence in nature and if the attempt to give the concepts of force, substance and causality a form which is based on such functional equations is successful the same will be true of these concepts. It is irrelevant whether or not these concepts, in the particular historical forms Mach attacks, are untenable; for our concern here is not with the results of particular investigations but quite simply with the question whether the existence of these concepts can be justified at all. And it must also be borne in mind that these are very plastic concepts and have not yet achieved their final form.

For this reason the idea we came across in the discussion of the concept of substance: that no real constancy corresponds in nature to just this concept, cannot be decisive. For once we examine the content of the objection we see that it amounts to the claim that what is captured by the concept of substance, once modern methods are applied, is not the persistence of a spatio-temporally individuated unity but that of a 'group' of functional dependences which occur as reactions 'here and there'; but there is absolutely no compulsion (Mach, at least, does not show that there is) to restrict the philosophical concept of substance to that form of the concept criticised by Mach. Once the brunt of the attack has been dealt with in this way its justification is further diminished when one considers that already in that constancy of reactions which Mach himself speaks about there is an indication of something which persists and in the constant cohesion and interrelation of a group of equations there is to be found an indication of a real moment[55] unifying the relations they express – the more so if, like Mach, one sees the essence of equations in a reproduction of the facts. Whatever more exact epistemological analysis makes of this real moment corresponding to the equations, and whatever changes this leads to in our ideas about such a moment, it may not be neglected. For to neglect it now is simply to avoid putting a question which is called for by the facts themselves.

And the same is true of the second fundamental concept under attack, causality; here too it is the facts themselves which direct us to the concept. For example, a certain quantity of work is functionally connected through the relevant equation with a particular quantity of heat. At the same time it is also a fact – albeit one which, though it is not expressed by the equation, definitely belongs in any discussion of it – that friction generates heat but heat does not generate friction except by a very different, indirect route. The presupposition that all such uni-directional connexions can be analysed into simultaneous, reversible connexions is for the time being a mere promissory note, a pipe-dream. But even if exact inquiry were to show the connexion to be what Mach assumes it to be in the example of the shot, this would still not exclude the possibility that causal relations obtain between those components of such processes which are precisely not immediately adjacent to one another. That is a matter belonging to a complete elucidation of causality. Mach himself mentions the fact that if two physical magnitudes are connected a change in one may correspond to a change in the other but that the reverse may not always be the case[56]. Changes in the values of physical magnitudes may under certain circumstances occur only in a single direction. "Of the two analytical possibilities only one is actual. We do not need to see in this fact a metaphysical problem", Mach says[57]. But there is no doubt that the fact that whereas here only one possibility is actual, where in other cases both are physically significant, contains something which points beyond mere functional dependence[58]. There is then in any case a factual basis to the concepts which have been eradicated so simply in this way. And it is this factual basis which Mach never sufficiently takes into account. He deals with equations as though they were merely aids to calculation, tools in the service of the economy of thought, and as we shall see more clearly in the next chapter, he deals with "merely logical" dependence as though it were something arbitrary. Concepts based on equations understood in this way do then of course appear to be without any objective foundation, mere temporary props to be put up and taken down as one thinks best. But to view matters in this way is to exaggerate the situation.

Or is the reference to universal connexion supposed to save the situation? All states depend on one another, we were told. Moreover, concepts presuppose equations and equations presuppose concepts. Is it, then, not likely that both are merely provisonal aids with which we single out certain nevertheless not completely separable moments from the general contexts[59]. This seems in fact to be the intention; a sort of πάνταῥεi. It played a role already in the previous chapter; but as we have already emphasized there and later, there are also in the general flux of phenomena very definite bases for the formation of certain concepts, concepts, to be sure, which are 'controllable by experience[60]' because they are built up on the basis of experience. To remain with the Heraclitean comparison: the flux of phenomena exhibits certain peculiarities in the way it flows which make possible the assumption of constant structures determining the direction taken, even if these are not immediately visible. Against this, Mach emphasizes considerations pointing in the direction of ever further dissolution. But however it is formulated, if the difficulties we have indicated at every step in Mach's argument are taken into account as well as the always available possibilities of alternative interpretations, then what Mach says turns out to contain objections, directives, clues but no rigorous demonstrations.

On the other hand, we too have had to limit ourselves to hints and comments. Nevertheless we have here two opposing views. Both believe they are guided by experience but one points left the other right. We must here forgo a refutation of Mach (through elaboration of the view opposed to his), for we do not want to begin with investigations of our own here, nor do we want to appeal to those of others which Mach perhaps does not even recognize. It only remains for us therefore to investigate the question whether Mach's point of view is, at least internally sufficiently well established and whether its development is at least free of contradiction.

The more thorough examination of the next chapter will show that this is not the case.

V The final component of the concept of 'functional connexion' completed: the denial of natural necessity. The theory of elements. Final contradictions.

We saw in the last chapter that Mach considers functional connexions exclusively from the point of view of cognitive grounds and consequences and that he also overlooks the fact that even a logical connexion can only provide a cognitive ground if justified by some objective foundation.

We shall now examine further this interpretation, which seemed to be necessary if Mach's attitude was to be understood, and we shall see that it is in fact correct.

Consider first the following remarks, which supplement the description of Mach's view on causality in the last chapter. His starting point is Hume's thesis that the basis of all causal judgements is to be found in habitual expectation. He goes on to add that we therefore judge concerning the connexion of two facts in very different ways according to circumstances: in some cases we scarcely think of the possibility of a connexion at all, in other cases we are under what is no less than a psychical compulsion – the connexion seems to us to be necessary. Mach interprets this as follows. A determinate trajectory seems, for example, to be necessarily connected with the initial velocity and direction of the projectile. The process is in fact given when it conforms to the well-known kinematic laws; then the initial velocity and direction become the cognitive basis yielding the elements of the trajectory as a logically necessary consequence. This logical necessity is, certainly, something which is felt, but it must also be borne in mind that it only obtains under the specified conditions and that the satisfaction of such a condition is merely given by experience without being due in the least to any sort of necessity[1].

This suffices to indicate just how exclusive are both Mach's emphasis on necessity as merely a matter of logical consequence and his refusal to recognize any other sort of necessity. And similarly we read: "If I find that a physical fact behaves like my calculation or construction, I cannot at the same time assume the opposite. Thus, I must expect the physical result with the same certainty with which I regard the result of the calculation or construction as correct. But this logical necessity is obviously to be distinguished from the necessity of the assumption of the parallelism between the physical fact and the calculation, this assumption being invariably founded upon a common experience of our senses. The strong expectation of a known result, which appears to the scientist as a necessity, is based upon the practice of firmly associating the conception of facts with that of the different aspects of their total behaviour . . . In this way there develops what is normally described as a 'feeling for causality'[2]".

And:

"A complex of physical facts is something simple, or can at least in many cases be experimentally arranged in such a simple form that the immediate relations between its parts become visible. Now if we have done enough work in this area to have enabled us to acquire, as regards the nature of these relations, conceptions which we think generally correspond to the facts, then we are bound as a matter of logical necessity to expect that any particular fact which may present itself will correspond to these conceptions. But this implies no necessity in nature. It is in this that 'causal' understanding consists[3]."

Finally, we find the direct claim: "There is only logical necessity: if fact A has certain properties this is not something I can simultaneously disregard. But the fact that it has these properties is simply given by experience. There is no such thing as physical necessity[4]."

The attempt to determine the meaning of these statements shows beyond all doubt that – as already mentioned – Mach holds necessity to be always logical necessity and that he apparently also takes this necessity to be psychological. For, we read, if A is found to exhibit behaviour B, if A is B and so on, then this cannot simultaneously be disregarded, it must necessarily be expected and so on[5]. It is less clear what led Mach to limit the scope of necessity in this way and to make this confusion; and just what sense he himself attaches to his claims.

We will, therefore, look first at a supplementary train of thought, the kernel of which is roughly this: "To explain means to resolve more complicated facts into as few and as simple ones as possible. These simplest facts . . . are always unintelligible in themselves, that is to say, they are not further resolvable. An example of this is the fact that one mass imparts an acceleration to another. Now it is only an economical question and a question of taste, at what unintelligibilities we stop. People usually deceive themselves in thinking that they have reduced the unintelligible to the intelligible. Understanding consists in analysis alone; and people usually reduce uncommon unintelligibilities to common ones. They always get, finally, to propositions of the form: if A is, B is, in other words to propositions which must follow from intuition and which, therefore, are not further intelligible[6]."

This means, then, that to explain, to understand, is no more than a process of analysis which yields what is simple, a process of resolution which yields what is familiar.

"It is always only a matter of recognizing the same elements in all facts, or, if one wishes, of finding those elements in a fact which are elements of another already familiar fact[7]", Mach says, for "when we survey a domain of facts for the first time, it appears to us diversified, irregular, confused, full of contradictions. We first succeed in grasping only each single fact, not the connexions between them. The domain is, as we say, obscure. Gradually we discover the simple, permanent elements of the mosaic, from which the whole domain can be mentally constructed. When we have reached the point where we recognize the same facts in the multiplicity we no longer feel lost in this domain; we survey it without effort; it is explained for us[8]." For "in fact, we consider a process explained when simpler, known processes are discovered in it[9]". "When experience has once clearly exhibited these facts and science has marshalled them in an order economically and perspicuously arranged, there is no doubt that we shall understand them. For there has never been any other sort of understanding than mental mastery of facts. Science does not create facts from facts, but simply orders known facts[10]." Mach relies in this connexion on examples from the natural sciences, some of which we shall now mention in order to see in what respect they support his position. Archimedes deduced his general principle of the lever (equality of the product of weight and lever-arm on both sides of the point of support as characteristic of equilibrium) from the proposition, which he took to be self-evident, that magnitudes of equal weight acting at equal distances from their point of support must be in equilibrium. Thus he considers the case of the lever to be explained "when simpler, known processes are discovered in it[11]". In fact even the simple proposition is not self-evident for such an assumption involves a great many presuppositions, for example that the position of the observer, other occurrences in the vicinity and so on exercise no influence[12]. It is only the expression of an experience, and an explanation in which it figures appears as a reduction to a fact which, though familiar, is as unintelligible as the derived fact. This is true even of the most comprehensive laws, from which entire domains of experience can be deduced, such as the principle of virtual displacements. It can be shown that they too contain nothing more than the recognition of some fundamental fact, in this case that particular natural occurrences take place, of themselves, only in a definite sense and not in the opposite sense[13] and, in particular, that heavy bodies, of themselves, move only downwards[14]. We have already considered the example of the explanation of planetary motion by the law of gravitation; it also belongs here. The problem is to explain the paths of the planets described by Kepler. Newton pictures their motion in terms of bodies swinging round at the end of a string; that the required tension in the string can be calculated enables him to establish that the planets move like heavy bodies whirling round at the end of a string with a certain tension, in other words, that the notion of terrestrial gravity can be applied to the planets too[15]. If, now, the tension of the string is replaced by the presupposition of tension alone or attraction between individual particles of mass as expressed in the well-known formula for gravitation this in no way alters the character of the connexion as a whole, which is that of the registration of a fact or, more exactly, that of the general description of a fact in terms of elements[16].

Mach comments further on examples of this sort:

"The proof of the correctness of a new rule can be attained by repeatedly applying it, by frequently comparing it with experience, by putting it to the test under the most diverse circumstances. This process would, in the natural course of events, be carried out in time. The discoverer, however, hastens to reach his goal more quickly. He compares the results that flow from his rule with all the different experience with which he is familiar, with all the older rules, repeatedly tested in times gone by, and watches to see if he does not light on contradictions . . .It is in this way that Archimedes proves his law of the lever, Stevinus his law of inclined pressure, Daniel Bernouilli the parallelogram of forces, Lagrange the principle of virtual displacements. Galileo alone is perfectly aware, with respect to the last-mentioned principle, that his new observation and perception are of equal rank with every former one – that it is derived from the same source in experience. He attempts no demonstration . . . It is quite in order, when a new discovery is made, to resort to all proper means to bring the new rule to the test. When, however, after the lapse of a reasonable period of time, it has been sufficiently often subjected to testing, it becomes science to recognize that any other proof than this has become quite needless; that there is no sense in considering a rule as the better established for being founded on others that have been reached by the very same method of observation, only earlier; that one well considered and tested observation is as good as another . . . In fact this mania for demonstration in science results in a rigour that is false and mistaken. Some propositions are held to be possessed of more certainty than others and even regarded as their necessary and incontestable foundation; whereas actually no higher degree, or perhaps not even so high a degree of certainty attaches to them. It is precisely that degree of certainty exact science aims to render clearly which is not attained in this way[17]." "How", Mach goes on to ask, "can the impression arise that an explanation is more effective than a description? When I show that a process, A, behaves like another process, B, which is better known to me, A thereby becomes more familiar to me; and the same is true if I show that A consists of the succession or juxtaposition of B, C, D with which I am already familiar[18]. But in this process one fact is merely replaced by another fact, one description by another description that is perhaps better known to me. The subject may thereby become more familiar to me, a simplification may result; but no change in essentials can take place . . . If we ask when it is that a fact is clear to us, the reply must be: when we are able to reproduce it in thought-operations that are perfectly simple and familiar to us."[19]

We want now to proceed immediately to a third train of thought, which belongs with the first two. Mach views laws of nature as mere tables of individual facts, rules of derivation, construction rules, compendious instructions for the memory, as the following series of striking statements should make clear.

"If all . . . facts, . . . knowledge of which we desire, were immediately accessible to us, no science would ever have arisen. Since the memory of the individual is limited, the material must be arranged. This is done by a rule of derivation which replaces the gigantic tables of correspondences. The former ('this rule of derivation, this formula, this law') has, now, absolutely no more real value than the aggregate of the individual facts. Its value lies merely in the convenience of its use; it has an economical value."[20]

And in exactly the same vein:

"When the anatomist in his quest for agreements and differences among the attributes of animals succeeds in obtaining increasingly refined classifications, the individual facts representing the ultimate terms of the system are nevertheless different enough to have to be noted singly . . . Physics, on the other hand, reveals to us wide domains of qualitatively homogenous facts, differing from one another only in the number of equal parts into which their characteristic marks or features are divisible, differing that is, only qualitatively . . . Here classification is so simple a task that it rarely impresses us as such, and even in the case of infinitely fine gradations in a continuum of facts the number system lies ready to follow as far as we wish to go. The coordinated facts are here extremely similar and closely related, as also are their descriptions, which consist in the determination of the numerical measures of one given set of characters from those of a different set by means of familiar mathematical operations – methods of derivation. Thus, here, what is common to all descriptions can be found so that a succinct, comprehensive description, or a rule for the construction of all the individual descriptions, can be specified – and this is just what we call law[21]."

"In regard to isolated facts, there is nothing to do but simply to keep them in mind. If, however, entire groups of interrelated facts are known, such that the two connected attributes A und B belonging to them each form a series whose terms differ only in the number of equal parts into which they may be resolved, then a more convenient survey (Übersicht) and mental representation may be obtained. The angles of incidence (A) as well as the angles of refraction (B) of a series of incident rays and the temperature-excesses (A) as well as the temperature-losses per minute (B) of cooling bodies may be resolved into equal parts; and to every term of series A corresponds a term of series B. A systematically arranged table can now facilitate the survey by assisting or replacing the memory. Quantitative investigation begins here; and it is, as one sees, a special case of qualitative inquiry, applicable only to series of facts which are related to one another in a particular way. We gain a new facility if the entire table can be replaced by a compendious rule for reconstruction, if, for instance, we can say: multiply the temperature-excess u of the cooling body by the coefficient μ and you obtain the temperature loss per minute u.μ." [22].

"It has already been mentioned that quantitative scientific statements are to be regarded as simpler and at the same time more comprehensive special cases of qualitative ones. Zinc, when acted upon by dilute sulphuric acid, gives a colourless solution; iron gives a pale bluish green one; copper gives a blue one; platinum gives none at all . . . If a gas is enclosed in a vessel provided with a manometer and a thermometer, I find, for different thermometric indications, different positions on the manometric column. I have here again a series of different cases which, however, have great similarity amongst themselves and differ only in the number of the thermometric degrees and the number of units of length of the manometric column. If I enter in a tabular form the position of the manometric column for every position of the thermometric one, I am then attending only to the schema of the aforesaid chemical arrangement. But I am at an advantage in that the thermometric and manometric positions each form a series between whose terms I can discriminate as finely as I please by a mere application of the number system, and without any further discovery. A further glance shows me that the separate cases represented in the table exhibit great similarity among themselves, that every position of the manometric column can be obtained from the thermometric position by a simple numerical operation, that this operation yields the right correlation for all cases in such a way that the whole table may be replaced and rendered unnecessary by the compendious rule for its construction,

$$p = p_o\left(1 + \frac{t}{273}\right)$$

etc.[23]"

"In more highly developed sciences rules for the reconstruction of great numbers of facts may be embodied in a single expression. Thus instead of noting individual cases of light-refraction we can mentally reconstruct all present and future cases if we know that the incident ray, the refracted ray, and the perpendicular lie in the same plane and that $\sin\alpha/\sin\beta = n$. Here, instead of the numberless cases of refraction in different combinations of matter and under all different angles of incidence, we have simply to note the rule above stated and its values – which is much easier. The economical purpose here is unmistakable. In nature there is no law of refraction, only different cases of refraction. The law of refraction is a concise, compendious rule, devised by us for the mental reconstruction of a fact, and only for its reconstruction in part, that is, in its geometrical aspects[24]." And Mach comments further on rules or instructions for constructing tables: "Practical needs require a familiar and sure application of science. This application is furthered by tracing back new relations to those already known . . . If, for practical reasons, one points out that fact A behaves in the same way as some fact which is familiar to us, B, then B may be a personal activity, an operation of calculation, or a geometrical construction. The spaces traversed by a falling body behave like the numbers obtained by squaring the numbers which measure the time elapsed; the temperatures of mixtures behave like arithmetical means, and so on. The more familiar such operations are to us and the simpler they are, the better we are satisfied, the less is the need for further explanation and the better do we understand the situation. All the peculiarity, certainty und familiarity of arithmetical operations carries over to the knowledge of the facts they represent[25]."

Let us now assess these three related groups of ideas; we have deliberately reproduced them in some detail. First, because they are of the greatest importance for the whole system, the epistemological foundations of which they must complete. Secondly, in order that with their help alone we might demonstrate their lack of clarity and their incompleteness.

The only thing which is really clear is, as we saw some time ago, that it is a psychological approach which everywhere makes itself felt. We hear of 'feeling' necessity, of strong, habitual 'expectations' which cannot be 'disregarded'; explanation is considered to be analysis into what is familiar because we are then no longer 'surprised', because reduced effort is then required of us if we are to 'survey' the area in question; clarity is attributed to a fact when it can be reproduced by quite simple and familiar thought operations, a proof is the equivalent of repeated acts of direct testing and so on. And in a law of nature all that is taken into account is really only the algebraic symbol, the rules of derivation and reconstruction "which is just what we call a law".

Now the psychological approach to all these matters is certainly legitimate and there can be no objection to Mach's observation that inquiry may well take this side of the matter as its starting-point[26]. But the question remains whether this is what is really meant here. And there can be no doubt about the fact that this is not what is meant, at least not the sense required by the context. For what we want to know, and indeed must know, is why there can exist *only* logical (=psychological) necessity, *only* classification instead of explanation, why *only* tables for us instead of laws about things. Only the demonstration of this claim that the one thing excludes the other can be of importance to us, whilst it is entirely irrelevant whether, apart from the objective, necessary connexion there are also involved subjective aspects and functions of the natural sciences. If one considers this question, then, although it is by no means easy to extract a definite sense from the remarks considered so far, the following are probably the only interpretations which need to be taken into account.

1. These remarks may mean that there is nothing in nature which corresponds adequately to our concept of necessity (a concept which, it should be added, is characteristic of inner perception alone and which, because of this, is a logical or psychological concept). This would mean that although we see connexions in nature which are as regular as necessary connexions would be, we lack any further insight into this natural necessity, every attempt to understand it is meaningless, especially if it means attempting to carry over just that concept of necessity which has its roots in our inner life. Where necessity is understood in this way, it is quite possible to say that there is something in nature which, in certain respects, looks like necessity but not that it is

necessity. – The resolution of causal relations into functional relations would be compatible with this interpretation, a resolution which in a sense exteriorizes the causal relation, since the functional relation itself is nothing more than the mathematical correlations that we make; although something must correspond to this in nature, no adequate correspondence is guaranteed. The view that laws say no more than a 'collection' of facts and merely reproduce these would also belong here. On this view, there is nothing over and above the facts which guides them and makes them necessary[27] or which is in any sense more easily understood than the facts. This would bring us back to the view that to explain facts by laws and deduce laws from more general laws is merely to connect one with another experiences which, taken by themselves, are all equally unintelligible.

2. The assertion that necessity exists only in the domain of the logical may refer merely to the degree of certainty involved and mean that we are not certain of any real necessity in nature since our assumptions, gained as they are on the basis of induction, are subject to error. Thus we read: "The agreement of concepts with one another is a logically necessary requirement, and this logical necessity is also the only necessity of which we have knowledge. The belief in a necessity in nature arises only where our concepts are closely enough adapted to nature to ensure a correspondence between the logical inference and the fact. But the assumption of an adequate adaptation of our ideas can be refuted at any moment by experience[28]."

3. Perhaps the direction taken by Mach's argument is this: that, strictly speaking, events in nature are subject neither to laws nor to rules and that they only appear to be subject to these to a limited extent (just that required for certain practical purposes).

The next step must now be to distinguish between those possible interpretations which are relevant to the systematic structure of the whole argument and those which can be ruled out as irrelevant. In view of the extensive evidence at our disposal this presents no great difficulty.

1. and 2. by themselves can immediately be seen to lack any significance for the system as a whole. For the fact that one cannot, so to speak, get a handle on necessity from the inside but only from the outside, that it is not necessity but only regularity which is perceived, and perceived not with certainty (Evidenz) but with a degree of probability which is capable of being increased, all this is merely a peculiarity of every empirical science and one which, though undeniable, has also never been

denied. Were Mach's opinion no more than this we would be obliged to describe as grave errors all those remarks considered so far which flatly claim that there is no such thing as natural necessity; Mach would then be inferring from the circumstance that there is a difference in the way necessity is grasped to the absence of any such thing as necessity. (And here we are ignoring the fact that he may not interpret logical necessity, originally postulated in opposition to natural necessity, as a psychological necessity – otherwise his argument would be circular, since the latter cannot be conceived except as a natural necessity.) If, then, we do not want to consider the stronger pronouncements to be misleading, we are left only with the stronger interpretations. We are faced with a choice; either what Mach says is unclear but essentially tame and at one with ordinary opinion or only the stronger interpretations are to be taken into account. There can be little room for doubt about the decision if one bears in mind the thrust of the whole argument so far and considers together its main divergences from the norm.

For what was the result of the third chapter? We considered two possible interpretations of the conceptual critique reproduced there: a careful, cautionary interpretation which requires that the connexion between scientific concept-formation and the facts of experience be as direct and intimate as possible; and a more radical interpretation according to which there is no access possible to what cannot be given immediately in sense experience and it is impossible to rise above the level of palpable experience. We said there that only the second interpretation should be taken as Mach's opinion not only in view of the structure of the whole argument but also because of the still to be discussed theory of sensualism in particular. At the same time we emphasized that even on this interpretation complete justification for this opinion of Mach's was still missing; the more so in view of the absence of any demonstration that it is possible to understand experience in a scientifically completely satisfactory way without the assumption, to which Mach is so hostile, that concepts transcend the sphere of what is perceivable. – In the fourth chapter we came to see the importance of functional connexions. We saw that they serve first and foremost to calculate the relation between quantitative attributes and that it is therefore a likely assumption (although not yet justified) that their significance is purely economical. But this step has to be taken if functional relations are to retain their relevance to the previous arguments and are to support these. For only then can it be asserted with some appearance of justification that because concepts are based on functional equations and their content is exhausted by the experiences expressed in these equations, this content itself is no more than a comprehensive, economical symbol for calculating certain experiences from certain other experiences. And as is easily seen, it is only this aspect of functional connexions as calculations which may be present in exact science if scientific concepts are not to contain what would otherwise give rise to the search for any further aspects; only if this is presupposed can the appeal to exact inquiry provide,at least in part, a foundation for the arguments presented above.

But this is of course by no means achieved merely by discarding the old, two-termed concept of cause – "a dose of effect follows on a dose of cause[29]" – or by emphasizing the role of calculation, but only by showing that the relations which find their expression in equations are not based on any necessity in nature. For as long as the equations express actual, law-like relations (in fact it would normally be presupposed that they only guarantee that calculation is possible if this is the case) they point to real, necessary connexions and it remains possible that the concepts they contain, rooted as they are in the behaviour they describe, have a real significance. Without the denial of natural necessity a hole would open up in the argument as a whole.

We did then in fact encounter a polemic against necessity and thus it is the context of the argument as a whole, more than particular passages, which compels us to interpret it as such, as directed against necessity. For this makes intelligible the interest in the dissolution of the concepts of force and substance; the notion of natural necessity is unavoidable so long as the belief in substances endowed with stable forces of their own persists, because this in its turn would be senseless without the asumption of real necessity. And finally, it is along these lines that the first objection to causality – that nature is singular and knows no repetition of the same cases – settles into place. We passed over this objection in the last chapter without discussion because we recognized that it is not only directed against causality but undermines absolutely all necessity and law-likeness in nature[30]. We see, then, that the whole argument comes to a head in the third of the interpretations we considered. This interpretation is necessary not only if certain pronouncements are not to begin to look like exaggerations, in the way we have already shown, but also if the whole edifice of Mach's ideas is not to collapse into a heap of unimportant and, we may say, careless statements. Conversely, we are now justified in letting everything turn on whether what Mach says, on this interpretation of his views, stands up to examination.

There is little doubt that necessity and law at least appear to exist in nature in the form of absolute regularities; it is from these that the presence of law-likeness in nature is derived even where this step is combined with the belief that nothing can be said about these regularities except that their existence is probable. Mach's attitude towards them is therefore of the greatest importance. "The business of physical science", he says, "is the abstract quantitative expression of facts[31]." But "every scientific proposition is an *abstractum* which has as its basis the recurrence of like cases[32]," for "in the reproduction of facts in thought, we never reproduce the facts in full, but only that side of them which is important to us; our reproductions are always abstractions[33]." Because "a rule, reached by the observation of facts, cannot possibly embrace the entire fact in its inexhaustible variety; on the contrary, it can only furnish a rough outline of the fact, one-sidedly emphasizing the feature that is of importance for the technical or scientific end in view...So, for instance, the weights and the lengths of the lever-arms were regarded at first as the conditions that determined equilibrium, then the statical moments etc. and finally the weights and the directions of the pulleys with respect to the axis were taken to be the conditions determining equilibrium and the enunciation of the rules modified accordingly.[34] In other words: "The progressive refinement of the laws of nature and the increasing restriction of expectations correspond to a more precise adaptation of thought to fact. It is of course not possible to achieve perfect adaptation to every individual and incalculable future fact. The extensive applicability of laws of nature, with the greatest possible generality, to actual concrete cases is only made possible by abstraction[35], by simplifying, schematizing and idealizing the facts: we must decompose the facts mentally, into such simple elements that from them we can mentally reconstruct and reassemble the facts with sufficient accuracy. Examples of such simple idealized factual elements, which never occur exactly in reality, are uniform and uniformly accelerated motions of masses, stationary (steady) thermal and electric currents, as well as uniformly increasing or decreasing currents and so on. Every arbitrarily variable motion and current may be regarded as made up to any degree of accuracy from such elements, so that the laws of nature can be applied to them. This occurs in the differential equations of physics. Our laws of nature thus consist of a series of theorems, appropriately chosen for this use and lying ready for application. Natural science may be viewed as a kind of collection of instruments for the intellectual completion of any partially given facts or for the restriction, as far as may be required, of expectations in future cases[36]."

The important new thought which appears in these remarks is the emphasis on the idealizing and hence fictitious aspect of natural laws. Our laws of nature, says Mach, are all gained by abstraction, by disregarding the full variety of facts. Only by idealizing the facts are we able to find laws. "All general physical concepts and laws, the concept of a ray, the laws of dioptrics, Mariotte's law and so on are obtained by idealization. This is what gives them that simple yet general, unspecific form which makes it possible to reconstruct any fact, even a complex one, in terms of synthetic combination of these concepts and laws and so to understand it. Such idealizations occur in Carnot's work: the absolute insulator, absolute isothermy of touching bodies, reversible processes; in Kirchhoff's notion of the perfect black body and so on[37]."

If this is the case and laws can only be found with the help of idealizing fictions then, Mach argues further, they presuppose that where the circumstances are the same the same events will be repeated. But because this presupposed regularity is a mere *abstractum* it cannot exist in nature, but only in abstraction, in an idealized schema.

And this brings us to the real misunderstanding on which the whole denial of natural necessity is based. Necessity, Mach concludes, is only to be found in the relations of mutual dependence between our concepts, in the ideas we have of law and so on; but since these are gained by idealizations, necessity can only be read into nature in a fictitious fashion. "For scientific purposes", says Mach, "our mental representations of sense experience must be submitted to conceptual formulation. *Only thus* may they be used for discovering by abstract mathematical rules unknown properties dependent on certain initial properties having definite and assignable arithmetic values; or for completing what has only been partly given. This formulation is effected by...idealizing[38]" for "it is only our schematic reproduction in thought that produces like cases. Only here does the reciprocal dependence of certain features exist[39]." Further, to this unambiguously determined dependency there corresponds "only a theory that represents the invariably complicated facts of observation, influenced as they are by numerous subsidiary circumstances, more *simply* and *precisely* than can really by guaranteed by observation[40]." It is only because "mathematical physics represents the objects of experience by means of schematizing and idealizing concepts that it can become an exact deductive science[41]." For an exact relation is yielded only by idealization and "appears...only as a hypothesis without which the individual facts of experience would at once become involved in logical contradictions. Only at this stage can we reconstruct the facts by operating with exact concepts and acquire a scientific and logical mastery of them. The lever and the inclined plane are as much self-created ideal objects of mechanics as triangles are ideal objects of geometry. These objects alone completely satisfy the logical demands which we make of them; the physical lever satisfies these demands only to the extent it approaches the ideal lever[42]."

In other words: "logical deductions from our concepts remain intact so long as we retain those concepts[43]", but "the facts are not compelled to conform to our thoughts[44]". Rather "our thoughts and expectations conform to other thoughts, namely to concepts that we have formed of the facts. If we assume that a fact corresponds exactly to our simple ideal concepts, then our expectations will agree with them and thus will be precisely determined. A proposition of natural science always has a merely hypothetical sense: if a fact A corresponds exactly to the concepts M, then the consequence B corresponds precisely to the concepts N; the two correspondences have the same degree of accuracy. Absolutely exact and perfectly precise and unambiguous determination of the consequences of a presupposition is to be found in natural science only as theory, it does not exist in sensible reality[45]."

It is on these remarks that the denial of natural necessity rests and it is here that everything else comes to a head. We described them as a misunderstanding. But before we draw the consequences from this and turn again to criticism, it is necessary to take into account one last idea: Mach's sensualism, the theory of elements.

The analysis of sensations (to use Mach's term) or theory of elements (to use one of its most important concepts) raises so many problems that a whole chapter would be required to deal with all of them. But we have already narrowed down our task in the introduction to the single most important point, the step which is supposed to justify Mach's claim that his views follow directly from the nature of exact inquiry. In its most important parts the connexion between the two is now familiar to us; it gives rise to a peculiar epistemological attitude and so we find Mach himself saying, in the preface to "The Analysis of Sensations": "What is attempted here is not the solution of all problems but an epistemological change of direction." This epistemological change of direction which, as our investigation so far seems to show, the theory of elements involves, denies the real necessity underlying natural laws as a result of the role of idealizations and fictions in natural science. Laws and theories are considered as a mere collection of instruments[46], the content of which is fictitious and without any meaning of its own, a device for producing easily surveyable inventories of facts[47]. This much, then, can be assumed, although we shall go into it in more detail later and see that what is most important in the analysis of sensations can be traced back to it. But it should also be noted that the dissolution into elements is itself a further step towards this final epistemological picture. For although, as we have already seen, the concepts of substance are expelled from the domain of what is considered worthy of scientific treatment, as long as there persists the belief in the physical and the psychical, an inner and an outer world this operation will never yield a definite result; the cause of the disease remains, so to speak, in the scientific organism.

Viewed in this light, the analysis of sensations turns out to contain three main groups of related ideas, leaving aside subsidiary considerations[48]:

First, the results of the natural sciences treat only of connexions between sensations, the world is therefore a world of sensations.

The following will suffice to illustrate this idea. We know that physics is a science of experience, of facts, or as Mach puts it "the only immediate source of knowledge in the natural sciences is sense perception"[49], the interpretation of even the most abstract equation also leads to perception, to an intuitive, sensible basis, or, in Mach's words, "all calculations, constructions etc. are only intermediate means to the attainment of this kind of intuition"[50]. Now there is of course an enormous difference between a science of experience and a science of sensations, but Mach thinks he can bridge the gap by inferring as follows: equations are based on measurements, measurements reduce to basic measures, normally length, mass and time; but mass and time turn out, as we have seen, to be measured in terms of length. "Consequently, measurement of lengths is the foundation of all measurements. But we do not measure mere space, we require a material standard of measurement, and with this the whole system of manifold sensations is brought back again...Thus although the equations only contain spatial numerical measurements the measurements themselves are also merely the ordering principle which tells us what the members of the series of sensible elements are out of which we have to construct our picture of the world"[51]. In other words: "The laws of nature are equations between the measurable elements of phenomena"[52], a "quantitative norm" of sense-presentation[53].

The first objection which will, I think, be made is that this 'quantitative norm' can only refer to sense-presentations in the most indirect fashion. For if one thing is clear, it is that the elements which equations deal with in physics are conceptual not sensory elements. Everything that has been said so far concerned conceptual elements and would be incompatible with any other sort of element – one need only remember that individual components of equations were found to be idealized and fictitious, and so cannot possibly be encountered in sensible reality. Finally, Mach himself expressly speaks of concepts. "For scientific purposes the reconstruction in thought of sense experience must be given conceptual form[54], and for the physicist concepts are instructions about how something is to be constructed[55]."

Thus if Mach's original remarks are to retain their validity, concepts must be regarded as having the role of an intermediary. And Mach does indeed hold that the physicist always operates with sensations because his concepts are based on these. Every experimental set-up which enables us to formulate a law or whose description is the basis of the definition of a concept[56] "is based on an almost unending series of sensations, particularly if we take into consideration the adjustment of the apparatus which must precede the actual experiment...Thus a physical concept means nothing but a definite kind of connexion of sensible elements[57]." Mach explains the fact that we nevertheless do not speak in just these terms as follows: "natural science makes us acquainted with only the firmest connexions of groups of elements. We may not begin by paying too much attention to the single constituents of these groups if we want to retain a comprehensible whole. Instead of equations between the primitive variables, physics gives us, as much the easier course, equations between functions of these variables. Physiological psychology teaches us how to separate the visible, the tangible, and the audible from bodies... Physiology further analyses the visible into light and space sensations; the first into colours, the last also into their component parts; it resolves noises into sounds, these into tones and so on. Unquestionably this analysis can be carried much further than it has been. It will be possible in the end to exhibit the common elements at the basis of very abstract but definite logical acts of like form...Physiology, in a word, will reveal to us the truly real elements of the world[58]." It must of course be added that such an "account can only indicate an ideal, whose gradual and approximate realization remains the task of future research. Finding out what the direct connexions between elements are", says Mach, "is so complex a task that it cannot be solved all at once[59]". The direction in which this clarification – the result of long and painstaking research – is to be expected, can of course only be surmised. To anticipate the result, or even to attempt to introduce it into any contemporary scientific investigation would be to do mythology not science[60]."

We can proceed immediately to the second group of objections since it involves the same ideas. What is given to us from bodies is, says Mach, the sensations they produce in us (to use a common mode of expression), sense-contents therefore, "colours, sounds, temperatures, pressures, spaces, times and so forth, connected with one another in manifold ways[61]." Why do we go from here to the assumption that there are things? Because, the answer runs, we have a need to unify and integrate[62], and because this need is met by the fact that "in the great mass of sensations the sum of those which remain constant compared with those which are variable is always so great, expecially when we take into account the continuity of the transition, that it appears to suffice for recognition of the body as the same[63]." "What is relatively fixed and constant stands out, engraves itself on the memory and expresses itself in language[64]." But to think that there must therefore be some actual 'permanent nucleus' behind the appearances, a thing which 'brings about[65]' the appearances is to make the mistake of overlooking the subjective and arbitrary nature of representation and to hypostatize it as as though it were objective[66]. The error in other words is to declare to be absolute what is really only a relative constancy which suffices to ground a mere subjective unification but not an objective unity in addition. The constancy of a particular complex of sensations is also subject to various conditions (our behaviour and relation to our surroundings), and this too contributes to its being merely relative. But because these conditions are under our control and are easily brought about they are not always appreciated, and bodies, as representatives of complexes of elements, are taken to be continuously present[67]. This can even occur in cases where the will alone is not enough to bring about these conditions or where it is completely impossible for the complexes in question to be made evident to the senses[68]. But if these mistakes are avoided then, conversely, all that can be said is that things or bodies are symbols in thought which sum up groups of sensations, symbols which do not exist outside our thought[69] because the disappearance of the sensations means that the nuclei, which are a contribution of thought, lose their entire sensory content[70]. "Not things but... what we ordinarily call sensations are the real elements of the world[71]"; "it is not bodies which produce sensations but complexes of elements which make up bodies[72]". And, as the final result: "The sensations connected with the different senses of a particular man as well as the sensations of different men are dependent on one another according to laws. It is in this that matter consists[73]".

It is not necessary to say much about these arguments; what Mach has to say enjoys some semblance of justification – although less obviously so in this sober summary than in the fluent presentation to be found in his own account. This is merely a consequence of the fact that he makes use of assumptions which are provisional, primitive and completely unclarified. What is a bundle or complex of sensations? What is a law-governed connexion between sensations? Before ideas such as these can be considered they must be made scientifically precise. But this just brings us back to the structure of the argument as a whole: the exact, scientific understanding of the behaviour of things is to be found in laws and we saw that Mach himself emphasized just this; thus the whole question comes down to the earlier question, to what extent the laws of nature are laws which hold between sensations. Only where it is possible to tackle these issues directly can this question be dealt with. And here only the indirect interpretation, the intermediate role of concepts, is relevant; for were Mach to see the nature of substance in some other law-governed connexion between the sensations of different senses of different people than in this one, then it would be necessary to indicate those natural laws of the external world which refer directly to sensations; ordinary physical laws do not do this, nor is it open to Mach to claim that they do so without inextricably involving himself in a flat contradiction with the other implications of his approach, mentioned on p. 71.

We are left with a third group of objections according to which it is misleading to set aside the distinction between one's own sensations and those of others as well as that between sensation and what is sensed in favour of one uniform sort of 'element' which does not really belong either to the inner or to the outer world.

Let us assume the laws of nature to be a quantitive norm governing sense-impressions and showing us which of these we need, and in what combination, in order to reproduce familiar facts. Red, green, extension, pressure, etc., would then be the elements of the external world insofar as they are perceived. But the usual view still distinguishes even here between the elements insofar as they are given to the senses, and their (perhaps unknowable) nature, which is independent of the subjective conditions of perception. Mach opposes this distinction, claiming that red, extension and the like are already, so to speak, elements in themselves and that their ambiguous position between the physical and the psychical is based only on a confusion and a change of perspectives.

He says: Let us call these elements A B C ...; our body forms a particular part of these and will be designated K L M ...; finally, the properly psychical elements, moods, memory-images, feelings, volitions[74] and the rest are to be designated $\alpha\beta\gamma$ ... Initially, it is assumed that these are not essentially different from presentations (*Vorstellungen*)[75]. The $\alpha\beta\gamma$ ... are then the same as the K L M ... and the A B C ... At the same time K L M ... are more intimately connected with $\alpha\beta\gamma$ ... and with A B C ... than they are with one another[76], since our psychic processes do not influence processes in the external world directly but they do depend on processes in the nerves, that is to say changes in K L M ...; similarly, A B C ... and K L M ... are interdependent, for it seems that "different A B C ... are associated with different K L M ...;" thus the appearance of a body to the right eye differs from its appearance to the left eye, with closed eyes it is invisible, and so on[76]. What is given is always elements in different relations of dependence. All dualism has its source in these differences and in nothing else. If one ignores K L M ... and only takes account of the connexions in A B C ... one is doing physics, if one takes into account the connexions between the two one is doing psychology, and may call A B C sensations. To the extent that all A B C ... can be so connected, all elements can be considered to be sensations[77]. And the ego, which is built up out of sensations[78], can embrace the whole world[79]. "It is therefore important for us to recognize that in all questions which can be intelligibly asked here...everything turns on taking into consideration different ultimate variables and different relations of dependence. That is the main point. Nothing will be changed in the actual facts or in the functional relations, whether we regard all the data as contents of consciousness, or as partially, or completely physical[80]". But what then is the status of the sensations which we also assume others to have? They are, Mach thinks, contributions of our own thoughts, additions which we make on the basis of analogy[81] and for functional reasons[82], because they are what first make the behaviour of others familiar to us[83]. "The presentations of the contents of consciousness of our fellow-men play for us the part of intermediate substitutions by means of which the behaviour of our fellow-men, the functional relation of K L M ... to A B C ..., insofar as by itself (physically) it would remain unexplained, becomes intelligible. [84]"

We have now assembled in a more or less complete fashion the relevant materials and our efforts to display the contradictions and errors they contain can now be brought to an end. The course of our argument began with the requirement of a demonstration that experience can be grasped in a scientifically satisfactory way yet without going beyond what is perceivable. And it took us from there to the interpretation of functional connexion as a matter of economy and calculation, from this to the denial of natural necessity. And from there in two different directions: on the one hand, to the role of idealization and of the process of abstraction which, we said, anticipating later arguments, could only misleadingly be held to be the foundation of idealization; and, on the other hand, to the view of science as a mere economical inventory and collection of instruments, a view which follows from the denial of natural necessity.

It was at this point that we introduced the theory of elements; but with what result? We were left with two lines of thought. First, Mach believed the gap between a science built up on the basis of perceptions and a science of perceptual contents could be bridged by the laws of nature, which are ultimately ordering principles directing us to the sense impressions with whose help we are to put together our picture of the world, equations between the measurable aspects of phenomena, quantitative norms governing sense impressions. If this is the case then the concern of science would be merely to mediate between phenomena; this task would constitute the beginning and end of all science; and should it perhaps occasionally involve an inescapable surplus of meaning over and beyond what is perceivable then this would in a sense be just irrelevant, no more than a subsidiary implication attaching to the algorithmic symbol. But we showed that this claim can only be taken in an indirect sense, for the elements which occur in laws are conceptual not sensible, as Mach indeed admits. We therefore looked in what Mach says about concepts for something capable of mediating between phenomena and found this explanation: although contemporary physics, as a matter of convenience, deals not with equations between the basic variables but with equations between magnitudes which are already functions of these, it is nevertheless the case that concepts signify only a particular way in which sensible elements hang together.

Quite apart from the fact that Mach (cf. p. 72) reserves strict proof of this assertion for a science of the future, while himself advancing only aphorisms (which, since they do not really admit of any discussion, we too were obliged to put on one side until some attempt to render them scientifically precise is forthcoming), there is an immediate and fundamental objection to his assertion. Every empirical concept, we have conceded, has its basis in experience; but the sense of this is distorted if it is taken to signify a connexion between sense-contents. For although it is of course true that perceptions consist of sense-contents, that perceptions lead to concepts and hence are, so to speak, unified under a concept, nevertheless a concept signifies something other than a collection of perceptions, as an attempt to display its meaning makes clear. If I attribute the property α, mass for example, to a body, if it exhibits the scientifically determined behaviour α I can do so only because of perceptions of α here and there. But no less necessary is the fact that α itself is independent of its being perceived, remains unchanged whoever perceives it and so on. This much, then, follows from the prevailing view, quite apart from the fact that perceptions which are directed to the same object by no means presuppose that the contents of the perceptions are the same.

Yet it is precisely this view which the last group of arguments is supposed to show to be erroneous. The distinction between the elements given to the senses (the elements as and to the extent that they are perceived), and their nature as objects which are independent of the subjective conditions of perception, is said to be a false distinction. Why? Because, we were told, it is based only on a change in the direction of the investigation, on differences in the sorts of functional combination posited. Elements occur only once and are neither physical nor psychical; it is only with respect to other elements that they may be either one or the other. The fact that they are described as psychical insofar as they concern one's own body and physical insofar as they concern other bodies is of no importance whatsoever, it may even be misleading and is certainly unnecessary. For the legitimate interest of the scientist is exhausted once he knows, in each case, how the elements behave with respect to one another, what sort of functional dependence exists, and so on.

The question now arises whether, as Mach thinks, phenomenal dualism is really only an external addition to what is actually given, or whether it is necessary to the latter. One thing is certain, and a matter of experience: the elements A B ... of which Mach speaks are always bound to the presence of K L M ... for where there is, for example, no retina, there is no colour, or this colour could not be a Machian element and would have to be something behind the content of the perception. If one were to investigate elements A B ... in their dependence on D E ... one would therefore not be able to abstract from K L M ...; every physical enquiry would remain a psychological enquiry. Mach himself claims that the basic equations have the form F(A B ..., K L ...) = 0[85]. In what ways, then, is it possible to disregard K L ...? It would make sense to do so from any sort of sceptical position, where considerations of economy would dictate how facts should be ordered and replace any attempt to read off an order immanent to the facts themselves. It would also make sense if one were able to demonstrate the independence of A B C ..., in other words the lack of influence of K L M ... according to the requirements of the law of induction. The first possibility, as has been mentioned, amounts to the invocation of economy and the whole train of argument derived from this – up to that denial of real necessity which was the point we arrived at above. The second possibility, on the other hand, would simply be the demonstration that the physical elements A B ... are related to one another quite independently of their being contents, a feature which ceases to be true of them as soon as the elements K L ... are set aside.

But matters are exactly the same when the sensations of other people are added in thought; for Mach says, as we have noted that these are added in thought on the basis of analogy just as the idea of a neurological process is added to one's own sensation. Elsewhere[86] he even compares this with the case where, after observing that a wire possesses all the properties of a conductor charged with an electric current, one infers a property which has not been observed. But if these cases are parallel then there can be no doubt either that one must assume, as something which is scientifically certain, that other people have a psychic life or that this certainty is also denied in quite unexceptionable and recognised cases of induction. Thus we have in fact the same alternative as before: either Mach involves himself in absurdities or the argument flows into those already noted.

As far as concerns the final point, the talk of the different domains on which the difference between perception and representation is supposed to be based, Mach's opinion is that this means only that some element is combined with various other elements; one must then either take the latter to include only clear cases of merely-psychical elements $\alpha\beta\gamma$ (such as the functions [87] of modern psychology) and assume that the difference between the two domains is to be located here since the difference is not to be found in A B ... whatever the extent of their combinations with one another. Or one must look for the difference not in what is connected, but in the mode of connexion. The first of the two cases, of course, provides a basis for dualism rather than Machian monism; but differences in the mode of connexion, since they are relevant only to the extent that they are law-governed, point to differences in the law-governed structures of the realms of the psychical and the physical and so lead, once again, to a separation of the two or, where they are not taken into account, we find we are back with the old point of view.

The basic ideas of "The Analysis of Sensations", then, lead either to contradictions or back to the point to which we were able to reduce everything else. With the discussion of this point our task will be concluded.

Our assumption was that Mach denies and dismisses the existence of necessity in nature. There are explicit statements to this effect and the sense and progressive articulation of the whole system requires as much insofar as it would dissolve into a series of contradictions were the different ideas deprived of this unifying perspective.

We need now only point out that it is precisely this idea, which holds everything else together, which involves Mach in contradiction and flatly contradicts his own scientific activity. "Indeed we cannot decide to inquire into a field unless we assume that it can be investigated, which presupposes constancies; for if not, what is there to investigate?". These[88] are Mach's own words and, since the point is an important one, we shall cite more of the same. "In embarking on our investigations we assume with good reason that dependence is constant...Past experience affords us this presupposition and every new success in inquiry reinforces our confidence in it[89]." "Only what is uniform and conformable to law can be described and conceptually represented[90]." "We have not been mistaken in postulating the uniformity of nature even if, because experience is inexhaustible, we shall never be able to prove that the postulate is absolutely applicable . . .; like any tool of science it will remain an ideal[91]." "The fact that we make predictions with the help of a law is proof of the adequacy of the degree of uniformity of our surroundings[92]." "I am convinced that in nature only that ocurs which can occur and only as much occurs as can occur and this in only one way[93]." "More accurate quantitative inquiry aims at determining facts as completely as possible and at unambiguous determination.[94]" "The unambiguous determination of certain properties of facts, properties which are important to us, by others which are more easily accessible, is, therefore, what is aimed at in science[95]." "In the course of inquiry every scientist is necessarily a theoretical determinist, even if he is concerned with mere probabilities ... The propositions of the calculus of probabilities hold only if chance events are regularities masked by complications[96]."

These are all pronouncements of Mach's and show that, at least in these passages, Mach presupposes constant, law-governed and hence, we would object, necessary relations in nature and that he takes them to be capable of investigation as far as is thought necessary. It is possible that these passages are mere slips on Mach's part but the objection applies to numerous other passages, which we have not explicitly mentioned, where Mach recognizes equally clearly that facts exhibit uniformity, regularity and unambiguous determinateness and that they can be predicted and investigated. And quite apart from the magnitude of these contradictions, without this presupposition the whole construction would lack its foundation. For although we were able to connect the denial of necessity with the fact that our conceptual grasp of natural laws involves idealization, no demonstration was provided that because of this there could be no necessity behind the laws of nature. Similarly, no demonstration is forthcoming – although it is just as urgently needed as in the previous case – that it is quite possible to arrive at an epistemological position in keeping with the results and requirements of exact research even when all the implications of the denial of natural necessity are properly developed (which is not the case).

A second possibility, however, is to take the pronouncements quoted above and related pronouncements to correspond to Mach's real views, in which case the denial of natural necessity could only rest on a misunderstanding.

And a misunderstanding does seem to have occurred. For what Mach emphasized in support of his views was that the exact dependence *expressed* by laws obtains only between conceptually intended objects and that these are idealized and cannot exist as such in the world of perception. One can therefore talk of fictions here, but one may not regard the dependence as arbitrary. For it is founded in experience. "It is first of all experience which tells us what dependence relations there are between different phenomena and only experience can tell us this[97]," and "it is indeed we who make our concepts but it does not follow that we make them in a completely arbitrary fashion[98]", Mach says. Experience teaches us to recognize the existence of astonishing regularities. Thus, this regularity, which is what allows us to infer necessity in the first place, lies in the facts and clearly cannot be removed from the facts by any idealization. Indeed an exact analysis of the process of induction – something which cannot be carried out here – shows that this regularity is the basis of every step in this process of idealization; the idealization has its motivation in the facts. It is therefore also wrong to say that necessity is imported into the facts by idealization. There is a sort of necessity of which this can be said, one which obtains only between idealized concepts and which Mach therefore calls a merely logical necessity, a hypothetical necessity with antecedents which can never be satisfied – the existence of a perfect gas, of a frictionless fluid and so on. This sort of necessity does of course presuppose idealization but it is not itself necessity proper and only involves necessity because of the existence of necessity proper in the facts themselves, whether or not we are ever able entirely to grasp their true structure with the means at our disposal.

Whatever the source of the denial of natural necessity, if Mach abandons it – and this was the second possibility we took as our hypothesis – the views described earlier lose their individual justification; laws will then no longer be mere tables; mathematical dependence can give way to the real dependence on which it is founded and economy of experience to inquiry; theoretical connexions can be more than ordering relations. Since there are two quite distinct types of law, physical and psychological, sensation and law can once again be distinguished; the separation of these two sorts of things, which are connected by relations of reciprocal, law-governed dependence, makes room once more for causality amongst other things and makes what Mach says on the matter wrong and misleading.

But one way or the other, whether one holds to the recognition of necessity or to those views which can only lead to its denial, in each case one comes up against a conflict in Mach's own views. Whatever course the problems touched on may take in the future, Mach has not provided a clear solution, a completely satisfying point of view on the basis of which future solutions might be arrived at. This is true, of course, only of the final metaphysical and epistemological results as these have been considered here. As is widely recognized Mach's writings are full of the most sparkling comments and the most fruitful suggestions. But consideration of these would not fall within the scope of our task.

Notes

Works by Mach referred to, with the abbreviations employed here:

- *Die Geschichte und die Wurzel des Satzes der Erhaltung der Arbeit* (Prague 1872) – E.d.A.
  History and Root of the Principle of the Conservation of Energy (translated and annotated by Ph.E.B. Jourdain), Chicago 1911 (C.E.).
- *Populärwissenschaftliche Vorlesungen*, 3rd Ed., Leipzig 1903. –P.V.
  Popular Scientific Lectures (translated by Th.J.McCormack), 5th Ed., Chicago 1943 (P.L.).
- *Die Analyse der Empfindungen und das Verhältnis des Physischen zum Psychischen*, references normally to 4th Ed., Jena 1903 – A.d.E.
  The Analysis of Sensations (translated by C. Williams & S. Waterlow), New York 1959 (A.S.).
- *Die Mechanik in ihrer Entwicklung*, 5th Ed., Leipzig 1904 – M.
  The Science of Mechanics: a Critical and Historical Account of its Development (translated by Th.J. McCormack), Illinois 1960 (M.).
- *Erkenntnis und Irrtum*, Leipzig 1905 – E.u.I.
  Knowledge and Error (translated by P. Foulkes and Th.J. McCormack), Dordrecht 1976 (K.&E.).
- *Die Prinzipien der Wärmelehre*, 2nd Ed., Leipzig 1900 – W.L. [An English translation of this work, edited by B.F. McGuiness, is due to appear. A translation of pp. 39–57 of the German edition is to be found as an appendix to B. Ellis, *Basic Concepts of Measurement*, C.U.P. 1968. (– *Tr.*)]

I. Introduction: Nature of our task

1 A.d.E., second edition, 21 (A.S.30), likewise: "This... view is the one which does justice to the present temporary state of knowledge as a whole in the most economical fashion, without any pretension to being a philosophy for all eternity." *loc.cit* 23 (32). "Whether I shall ever succeed in making my fundamental ideas plausible to philosophers, I must leave to time to decide. I do not attach much importance to this at present, though I have a deep reverence for the gigantic intellectual labours of the great philosophers of all ages. But I have an honest and lively desire for an understanding with natural scientists, and I consider that such an understanding is attainable. I should like scientists to realize that my view eliminates all metaphysical questions, and in a way which is indifferent as to whether they be only regarded as insoluble at the present moment, or whether they be regarded as

meaningless for all time " A.d.E. 287 (A.S. 368–369). "...Our object is not to create a new philosophy or metaphysics, but to promote the efforts now being made by the positive sciences towards mutual accommodation." A.d.E. 259 (A.S. 332). "Above all, there is no Machian philosophy, but at best a scientific methodology and cognitive psychology, and both are provisional, imperfect attempts, like all scientific theories. I am not responsible for a philosophy that might be constructed from this with the help of alien additions." E. u. I. Preface (K. & E.xxxiii). "My expositions always start from physical details and from there rise towards more general considerations." E. u. I. 141 (K. & E. 103).

2 M. first edition, Preface (M.xxii).

3 E. u. I., Preface (K. & E.xxxii).

II. *The cognitive-psychological and economic approach*

1 Cf. A.d.E. 245 (A.S. 327–328).

2 E.d.A. 30–31 (C.E. 54–55).

3 Cf. E.u.I. 83, 164–165; A.d.E. 244; M. 7.

4 E.u.I. 443 (K. & E 352). Cf. E.u.I. 229; A.d.E. 246; M. 210.

5 W.L. 365.

6 E.u.I. 107, 110 (K. & E. 80, 81) cf. P.V. 218/219 (P.L. 189/190).: "The homely beginnings of science will best reveal to us its simple unchangeable character. Man acquires his first knowledge of nature half-consciously and automatically, from an instinctive habit of mimicking and forecasting facts in thought, of supplementing sluggish experience with the swift wings of thought, at first only for his material welfare. When he hears a noise in the underbrush he constructs there, just as the animal does, the enemy which he fears; when he sees a certain peel he forms mentally the image of the fruit which he is in search of; just as we mentally associate a certain kind of matter with a certain line in the spectrum or an electric spark with the friction of a piece of glass... These primitive psychical functions are rooted in the economy of our organism no less firmly than are motion and digestion... Such primitive acts of knowledge still constitute the most solid foundations of scientific thought even today." Cf. also: E.u.I. 2, 58, 182f., 229, 257; W.L. 120, 365, 386; A.d.E. 41, 246.

7 "Knowledge is a product of organic nature... and the general imprint of evolution and transformation must be noticeable in ideas also." P.V. 247 (P.L. 217/218), also W.L. 382.

8 Thus a mammal living in water does not acquire new extremities, there is rather a transformation of those it already has. P.V. 256 (P.L. 229), W.L. 388.

9 E.u.I. 126, 134/135 (K. & E. 93, 98/99) Together with this economic role of conceptual classification, mention should be made of the development of language and writing, which are closely connected with this role, and of their labour saving function. "Just as the different calls and cries of gregarious animals – mating calls, warning cries and battle cries – are unconsciously formed signs for a common observation or action, irrespective of the variety of possible occasions for such action, so too the words of human language are names or signs for universally known facts, which all can observe or have observed." P.V. 265/266, (P.L. 238). Cf. P.V. 220/221, 226, W.L. 119, 396, 412, 414, M. 522.

10 E.u.I. 110 (K. & E. 81), 134, 162, 298; A.d.E. 41, 248; P.V. 74/75, 245; M. 139, 531; W.L. 380, 381.

11 W.L. 394

12 E.u.I. 162 (K. & E. 120)

13 E.u.I. 162 (K. & E. 120)

14 E.u.I. 3 (K. & E. 2)

15 P.V. 253/254 (P.L. 225/226)

16 P.V. 276f. (P.L. 249f.) cf. on (1) and (2) E.u.I. 162–179 (K. & E. 120–147)

17 P.V. 223/224 (P.L. 194/195), cf. E.u.I. 190, 162–179, especially 171–174, 263f., 99, 230, 282–298 especially 284, 257; A.d.E. 249, 255; M. 27, 139, 196; W.L. 151, 385ff., 402; P.V. 226, 253/254.

18 cf.E.u.I. 243f. (K. & E. 176f.), 313; P.V. 256 (P.L. 229). Cf. also E.u.I. 109 (K. & E. 82); P.V. 256; W.L. 387.

19 P.V. 224 (P.L. 195); M. 526f. (M. 485f).

20 P.V. 257f. (P.L. 230f.); E.u.I. 180, 185f., 188; W.L. 388.

21 P.V. 258 (P.L. 230/231).

22 The formation of hypotheses mentioned above is of course also preceded and sustained by comparison. In addition, even elementary judgements and their communication are based on a process of comparison – (W.L. 396, 397; A.d.E. 248/249; P.V. 266 (P.L. 238); E.u.I. 240; M. 5, 6; and the most highly developed stages of science, too, lead to theoretical structures whose sinews are just such comparisons, analogies of enormous scope – E.u.I. 217–229, (K. & E. 162–170). Thus it is comparison which is "the most powerful, inner, vital element of science" and with respect to which one could even speak of 'comparative physics' – P.V. 266 (P.L. 238/239); cf. W.L. 396–406.

23 "In fact, all the varied methods of scientific inquiry ... enumerated by John Stuart Mill... are ultimately recognizable as forms... of the method of variation," we read at P.V. 257 (P.L. 230).

24 P.V. 222 (P.L. 192/193); M. 68, 131, 147; E.u.I. 201/202.

25 E.u.I. 441 (K. & E. 449).

26 M. 315, 382, 507, 520, 530.

27 P.V. 226, 279, (P.L. 197); E.u.I. 112, 127; W.L. 417/418; M. 192.

28 P.V. 232/233 (P.L. 204).

29 P.V. 226 (P.L. 197).

30 M. 549 (M. 504).

31 "The greatest advances in science have always been the result of successful formulation, in communicable terms, of what was instinctively known long before" P.V. 220 (P.L. 191), cf. P.V. 218/219; M. 180, 210.

32 For an example of this phenomenon of theories giving rise to other theories cf. P.V. 276f. (P.L. 249f.).

33 Cf. M. 272/273, 28.

34 On the influence of chance cf. W.L. 440–444. And on the specific example of the significance of the fact that Coulomb's torsion balance was constructed before Riess' thermometer cf. P.V. 198f., (P.L. 168f.); W.L. 322f.; E.d.A. (C.E.).

35 E.u.I. Preface

36 A few examples only: What knowledge is, is something we determine at our leisure (*loc. cit.* 5). There is no absolute, unconditional knowledge, only relative knowledge (*loc. cit.* 6). It is evident "from logic" that no assertion can have unconditional validity (*loc. cit.* 7). The only attainable goal of all science is subjective conviction not objective certainty (*loc. cit.* 9) etc. etc.

37 Compare for example A.d.E. 30 (A.S. 37): "No standpoint has absolute, permanent validity. Each has importance only for some given end." Or E.u.I. 114 (K. & E.) "Only success decides between knowledge and error." Compare too A.d.E. 257/8 and P.V. 235 (P.L. 206). In addition, where the replacement of the notion of a thing by the theory of elements is at issue, very sceptical passages are to be found. We will discuss these when we come to the position they occupy in the system; we will then be able to clarify their sense much more precisely than is possible here.

38 If one reflects on what follows from the remarks on pp. 24–25 above, it is clear that it comes simply to this: actual evolution, guided by economic and biological necessity, leads to very different ends, depending on circumstances. If this is borne in mind, it follows immediately that the sceptical turn of the argument to which it gives rise is unmotivated. For the courses

taken by actual evolution may be many and different; there is no doubt about this. But since the concept of an evolution which leads to correct results is a very different concept it remains untouched by the remarks above. Only the assertion that even results which are otherwise held to conflict with one another nevertheless enjoy equal justification would signify a restriction and – were this to be supported only by appeal to the diversity of actual evolution according to different circumstances – would contain the assumption that, in judging science, one can do no more than try to understand what has actually happened and that no other criteria are available. But it is precisely such a demonstration which is, we find, missing.

39 E.u.I. 162.

40 A.d.E. 285 (A.S. 365).

41 Thus the hypothetical pictures of mechanical physics which Mach opposes provide examples of the transfer of ideas corresponding to the principle of continuity. Cf.P.V. 187, 203; M. 562; W.L. 316f. (At one point – M. 532 [M. 588] – this is contradicted and atomism is described as a violation of the requirement of continuity; but this only shows the inner precariousness of the principle.) And elsewhere too the permanence of received ideas turns out to be an obstacle to scientific progress; cf.P.V. 167, 257, 269, 271. W.L. 21, 36/37.

42 Cf.W.L. 452/453; E.u.I. 446, 449/50.

43 A.d.E. 48.

44 W.L. 393/394.

45 W.L. 394; M. 530 (M. 586).

46 E.u.I. 176 (K. & E. 129).

47 E.u.I. 174 (K. & E. 127/8); cf.W.L. 391.

48 E.u.I. 282 (K. & E. 212).

49 Cf.E.u.I. 446 (K. & E. 354–5) where Mach first expressly demands univocity and only then economic ordering.

50 M. 537 (M. 592) Cf. also E.u.I. 282 (K. & E. 212): "General and indeterminate as this characterization of enquiry [as economical etc.] may seem, it is likely to contribute more to an understanding of the enquirer's activity than more specialized and therefore more one-sided accounts of it."

III. The opposition to mechanical physics. Criticisms of individual physical concepts.

1 E.u.I. 141 (K. & E. 103). Cf.M 537 (M 592)

2 The word 'Erscheinung' will be used here and elsewhere, unless otherwise indicated not in the pregnant signification it has of 'sense-content' [Sinnesinhalt], but in the sense in which one distinguishes between physical and chemical or electrical and magnetic phenomena. It signifies therefore the results of observation which form the basis of science.

3 E.u.I. 235 (K. & E. 174–175)

4 F. Rosenberger, *Die Geschichte der Physik in Grundzügen,* Braunschweig, 1882, II, 236/237. [*Sir Isaac Newton's Mathematical Principles of Natural Philosophy and his System of the World,* p. 547, trans. A. Motte, revised by F. Cajori, Cambridge U.P. 1934.] Cf. M. ch II, 3. E.u.I. 233 (K. & E. 173).

5 Rosenberger loc.cit. II 201. E.u.I. 235 (K. & E. 175)

6 In favour of the latter there is his pointed declaration *hypotheses pro nihilo sunt habendae.* Cf.E.u.I. (K. & E.) loc.cit. In favour of the former, the ideas he expresses in his correspondence with Bentley (cf.E.u.I. 234; M. 200), the fact that even his immediate pupils considered *actio in distans* to be a property of matter (cf. Rosenberger loc.cit. II,

237), the "disturbing narrow-mindedness" of his attacks on the wave theory (cf.P.V. 255), which suggests that he considered the emission theory of light to be more than a mere illustrative aid and similar considerations.

7 Cf.P.V. 185 (P.L. 155/156).

8 This was also the case before Huygens and Newton. Gilbert (1540–1603) introduced the notion of magnetic fluids and even Galileo made use of Aristotelian and atomistic ideas in certain cases. (Cf. Rosenberger *loc.cit.* II, 32, and E. Goldbeck, *Galileis Atomistik,* Bibl. math., third series, vol. III book 1).

9 Cf. for example E.u.I. 104 (K. & E. 77) where Mach pokes fun at the 'witches' sabbath' of atoms, ions, electrons, vortices, matter etc. Also A.d.E. 242 (A.S. 311), where the fact that the scientist is no longer overawed by the traditional intellectual implements of physics is described as a step forward; A.d.E. 26 (A.S. 334), where knowledge of the psychological genesis of such ideas is said to imply that they have only a relative value; W.L. 317, where they are described as very artificial, and elsewhere.

10 This refers mainly to the theory that heat is essentially motion and not, as was earlier taken to be, a caloric or stuff. Mach shows that the reasons on which this decision is based are inadequate. They owe their apparent justification to the accidental circumstance that when the measure of the quantity of heat was being established the choice fell on what later became important as its work-value; although this of course means that the quantity of heat disappears in the case of certain processes (where work is done) one cannot conclude from this, as is often done, that heat cannot be a stuff because its quantity does not remain constant, for the existence of this phenomenon depends on the measure chosen and this depends merely on historical factors. Mach goes on to illustrate this with reference to a similar situation in the theory of electricity where one can equally well arrive at conflicting results about the nature of the motive force. And, finally, Mach shows how even in the theory of heat a return to the conception of a heat-stuff is possible in view of the current situation there. This does not, of course, seem to him to be any more necessary than it did to J. Black for whom "Nearly every hypothesis can be made to agree with the phenomena by a skilful use of certain conditions, a fact which is agreeable to the imagination but does not improve our knowledge". [Translated from the German – *Tr.*] Cf.E.d.A. (C.E.); P.V. 196–201 (P.L. 166–171; W.L. 321–324, 179f.

11 The line of thought reproduced in the previous note continues as follows [E.d.A. 24, 27 (C.E. 47, 50) – *Tr.*]: "It is a matter of complete indifference and without the least scientific value whether we think of heat as a stuff or substance or not... But let us suppose for a moment that all physical events can be reduced to movements of molecules. What are we to do with the supposition? We would be assuming that things which can never be seen or touched and only exist in our imagination and understanding can have the properties and relations only of things which can be touched. We impose on the creations of thought the limitations of the visible and tangible. But why do we not think of molecular movements musically? It could even be advantageous to think of chemical processes in a space of more than three dimensions. Hence it is possible to hold on to the results of science independently of the mechanical theory, which is therefore unnecessary and often even a hindrance." Cf.P.V. 189 (P.L. 159): "After all, do we really know more about why a body leaves one place and appears in another than about why a cold body grows warm?" and what is said at W.L. 215/216 in the course of a discussion of Carnot's achievements.

12 P.V. 268, (P.L. 240); cf.E.u.I. 231f.

13 P.V. 268 (P.L. 240); cf.E.u.I. 231f.

14 P.V. 269 (P.L. 241).

15 P.V. 267ff. (P.L. 240ff) Apart from this heuristic value, hypotheses also have the advantage of providing a unified representation, as was mentioned above. "The advantage and scientific value of this idea lie in the fact that it reproduces in an intuitive and natural fashion the different facts observation has gradually and with difficulty brought together." P.V.

141. Cf. on these two functions also: P.V. 138; E.u.I. 223ff., 229–247, (K. & E. 167ff., 171–184); W.L. 123, 318. The hypothetical pictures of mechanics in particular have the advantage that they draw upon very familiar ideas which have already received thorough theoretical elaboration. P.V. 187 (P.L. 157), 203; M. 552; W.L. 316f.

16 E.u.I. 241f. (K. & E. 179f.)

17 P.V. 269, 257 (P.L. 241/2, 230)

18 Thus Huygens, caught up in the analogy with sound, cannot understand polarisation, P.V. 269 (P.L. 242). Black's successors were unable to profit from or appreciate the fact that heat is produced by fiction because of the conception of heat as a stuff, P.V. 167, 271, (P.L. 138, 244). Dalton burdens his works with questionable hypotheses, W.L. 21, 36/37 and so on.

19 E.u.I. 223f. (K. & E. 167f.).

20 P.V. 275, 267 (P.L. 248, 240); E.u.I. 244/245.

21 P.V. 272, 196/197 (P.L. 244, 166/167); W.L. 185/186, 193.

22 P.V. 192 (P.L. 161).

23 E.u.I. 267 (K. & E. 198/199).

24 E.u.I. 244 (K. & E. 181).

25 E.u.I. 244/255 (K. & E. 181).

26 P.V. 267 (P.L. 240. E.u.I. 244/245 (K. & E. 181): "The views which have arisen in this way are no longer hypotheses, but presuppositions of the intelligibility of facts and results of analytic investigation. These we can retain as certain, even if we can find no analogy at all for them..."

27 P.V. 275 (P.L. 248).

28 P.V. 235/236 (P.L. 206/207).

29 The correspondence need not involve perceivable features, it may be conceptual and consist of correspondence between conceptual relations. Mach defines analogy as a relation between systems of concepts in which the dissimilarity of two homologous concepts as well as the agreement in logical relations of every two homologous pairs of concepts is brought to light. E.u.I. 217, 218 (K. & E. 162, 163); P.V. 277 (P.L. 250).

30 E.u.I. 246 (K. & E. 182).

31 E.u.I. 227 (K. & E. 169).

32 Cf. P.V. 277 (P.L. 250); E.u.I. 226/227.

33 As at P.V. 267, 275 (P.L. 240, 248); E.u.I. 244/245 (K. & E. 181).

34 As at P.V. 277 (P.L. 250); E.u.I. 226/227.

35 E.u.I. 312 (K. & E. 234/235); A.d.E. 263.

36 The remarks on these subjects are to be found in the writings on the natural sciences and cannot be taken out of their contexts without misrepresentation. The references in what follows should therefore be supplemented by comparison with: on the concept of energy, E.d.A.; W.L., in particular 315–347, but important remarks are to be found elsewhere in the text; M., particularly ch.III; P.V., particularly XII. For the concepts of mass, inertia, space, time and motion, M., particularly ch.II. For the concept of temperature and quantity of heat, W.L., particularly 39–58, 153–195, and 211–347. For the fundamental concepts of electrostatics, P.V.XI and XII.

37 W.L. 315/316; E.d.A.; P.V. 208–214.

38 W.L. 321.

39 P.V. 205, 213 (P.L. 174); W.L. 340. Cf. p.XX, note 2.

40 W.L. 343.

41 Cf. the quotation, which has already been frequently mentioned, W.L. 321–324, P.V. 196ff. (P.L. 166).

42 W.L. 39ff.

43 W.L. 46, 48, 63.

44 M. 238 (M. 273). Thus the concept "acceleration of a freely-falling body at 9.810 metres per second" means that the velocity of the body with respect to the centre of the earth is 9.810

metres greater when the earth has performed an additional 86400th part of its rotation, something which itself can only be recognised through its relation to other celestial bodies.

45 M 242/243 (M. 277/279).

46 M 252f. (M. 283f.) i.e. between motion relative to the fixed stars on the one hand and, on the other hand, motion relative to other bodies held fast in space relative to the fixed stars.

47 M. 243/244 (M. 279/280).

48 M. 243 (M. 280).

49 M. 237 (M. 272–273).

50 M. 238 (M. 273).

51 M. 238 (M 273).

52 M. 243/244 (M. 280).

53 M. 230–270 (M. 264–305).

54 M. 231 (M. 266).

55 M. 233 (M. 268).

56 M. 232 (M. 267).

57 M. 236 (M. 271).

58 M. 143, 247ff. (M. 172, 288f.). It should be pointed out here that attempts have been made to see in the principle of inertia a natural law which is deducible a priori. Mach points out against this that the opposite of the law of inertia could be inferred with the same apparent justification provided one invokes only the general 'cessante causa cessat effectus'; it all turns on what one takes to be an 'effectus', velocity or acceleration, M. 143 (M. 172). We mention this here because of the assertion of the uselessness of causal considerations in physics, which will be discussed later.

59 M. 268 (M. 303). [The full text of Mach's 'experimental proposition', to which Musil here refers, is: "Bodies set opposite each other induce in each other, under certain conditions to be specified by experimental physics, contrary accelerations in the direction of their line of junction." (The principle of inertia is included in this.) – *Tr.*]

60 M. 140 (M. 169).

61 M. 247 (M. 284) i.e. behaviour in apparently absolute space.

62 M. 250f. (M. 287f.).

63 E.u.I. 112 (K. & E. 83).

64 P.V. 232/233 (P.L. 204). We will cite further remarks only after discussing Mach's attitude to a number of other important scientific concepts.

65 Here it is only fair to grant that the demonstration succeeds since we cannot go into possible criticisms; it should, however, be mentioned that discussion of the matter in specialist circles cannot yet be considered concluded.

66 Mach indeed makes just this point: "Faithful adherence to the method that led the greatest investigators of nature to their great results restricts physics to the expression of actual facts, and forbids the construction of hypotheses behind the facts, where nothing tangible and verifiable is found. If this is done, only the simple connexion of the motions of masses, of changes of temperature, of changes in the values of the potential function, of chemical changes, and so forth is to be ascertained, and nothing is to be imagined along with these elements except the physical attributes or characteristics directly or indirectly given by observation." M. 541 (M. 597).

67 Compare also note 15 p. 85 where the efforts of mechanical physics are rejected because the processes it bases itself on cannot be perceived.

68 This of course leads to the question, which cannot be dealt with yet, when something can count as being unquestionably gained from experience. Mach, it may be noted, is not consistent on this point. Thus there are passages where even the case we have chosen as an example, in which the vibrating of sounding bodies is inferred, is justified only in practical terms and not in terms of objective support which would put it beyond doubt. Cf. M. 531/532 (M. 587/588). When we come to consider the theory of induction which can be found in

Mach's writings, we will see that it contradicts Mach's attitude on this point. It is this theory which, quite generally, shows the necessity of going beyond experience and justifies doing so.

69 M. 237, 238, 244 (M. 273, 274).

70 Cf. A.d.E. 274 (A.S. 350): "When we think of excess of temperature as determined by the space traversed by a falling body, the dependence is not an immediate one... But the dependence is no more immediate when we assume excess of temperature to be determined by the angle of rotation of the earth. For no one will believe that the same temperature-values would continue to correspond to the same angular values, if the earth were to alter its velocity of rotation in consequence of some shock." But it seems to me that it follows from precisely such considerations as these that our postulates are merely provisional and depend on partial ignorance of the decisive part played by certain independent variables which are inaccessible to us. – Strangely enough, I am not the only person to draw this conclusion; Mach does so too *loc.cit.* But in his case this view is not compatible with the assertion that it is nevertheless the case that every attempt to go beyond this 'provisional', incomplete experience is senseless. Cf. also M. 261 (M. 297).

71 Here and elsewhere, e.g. pp. 49 ff., 54 f., Musil mentions an account of substance which is an alternative to Mach's account. According to this account the connexions or dependences between bodies and their individual reactions or spatial and temporal behaviour provide a basis in experience for the formation of a concept of property. On such an account see Stumpf, *Erscheinungen und psychische Funktionen,* 1906 and *Erkenntnislehre,* 1939/40, § 3. – *Tr.*

IV. The polemic against the concept of causality; its replacement by the concept of function

1 'Über die Erhaltung der Kraft' 1847.

2 Cf. E.u.I. 272 (K. & E. 204).

3 M. 524 (M. 580); cf. P.V. 228.

4 E.u.I. 272 (K. & E. 204).

5 A.d.E. (75 A.S. 92).

6 E.u.I. 273 (K. & E. 205).

7 A.d.E. 74 (A.S. 89); E.u.I. 273.

8 E.u.I. 273/274 (K. & E. 205/206).

9 E.u.I. 273/274 (K. & E. 205/206).

10 A.d.E. 74 (A.S. 89–90). The definition of mass, cited above, should be borne in mind here.

11 A.d.E. 75 (A.S. 91).

12 M. 524 (M. 580).

13 E.u.I. 274/275 (K. & E. 206.) Cf. A.d.E. 74.

14 The phenomenalism which is beginning to make itself felt here will be disregarded for the time being. We may substitute for the above: 'as soon as we can characterise the elementary determinants of events conceptually by means of measurable quantities'. In fact we shall see that this is the only possible meaning of Mach's assertion that there is a functional connexion between elements, although this is not always what he has in mind.

15 "The principal advantage for me of the notion of function over that of cause lies in the fact that the former forces us to greater accuracy of expression, and that it is free of the incompleteness, indefiniteness and one-sidedness of the latter. The notion of cause is, in fact, a primitive and provisional makeshift." A.d.E. 75 (A.S. 92). Cf. P.V. 281; W.L. 435/436 ; E.u.I. 273, 277.

[16] For an example, E.u.I.133.

[17] E.u.I.273 (K.& E.205). Cf. on the advantages of the concept of function and its position as the result of the actual development of science: A.d.E.74-78,80,262-264 (A.S.89-93,95, 336-338); E.d.A.35f.; M.282f (M.320f.).

[18] P.V.281 (P.L.254); W.L.435/436; A.d.E.74.

[19] A.d.E. 262-264 (A.S.336-338).

[20] Cf.E.u.I.273/274 (K. & E. 205/206); A.d.E. 262/264 (A.S. 336/338); M. 282f. (M. 320 f).

[21] M. 521 (M. 577).

[22] E.u.I.3.

[23] P.V.232/233(P.L.204).

[24] A.d.E.245 (A.S.315).

[25] Of course, these remarks are intended only to clarify Mach's point of view; they do not contain the position of the author.

[26] W.L.436/437, cf. W.L.379.

[27] M.232 (M.267).

[28] M.139f. (M.95f.).

[29] M.270f. (M.307f.).

[30] A.d.E.259 (A.S.332);cf.M.84;E.u.I.104;W.L.400.

[31] A.d.E.256 (A.S.328).

[32] A.d.E.258 (A.S.331).

[33] E.u.I.133f.(K.& E.98).

[34] E.u.I.134 (K.& E.99).

[35] E.u.I.277 (K.& E.207-208).

[36] "The physicist who sees a body flexed, stretched, melted and vapourised cuts up this body into smaller permanent parts; the chemist splits it into elements. Yet even an element such as sodium is not unalterable. When warmed, the white, silvery mass becomes a liquid which, when the heat is increased and the air shut out, is transformed into a violet vapour, and on the heat being still more increased, glows with a yellow light. If the vapour condenses, the white metal reappears. Indeed even after the metal has been brought into contact with water and has turned into sodium hydroxide, properties which have vanished completely can be made to reappear by suitable treatment; just as a moving body which has passed behind a column and is lost to view for a moment may make its appearance after a time. It is unquestionably very convenient always to have ready the name and thought for a group of properties wherever these may occur. But such a name and thought are no more than a compendious, economical symbol for these phenomena." P.V. 231 (P.L.202)

[37] M.523 (M.579).

[38] P.V.229 (P.L.200).

[39] P.V.232 (P.L.204).

[40] In order to complete the picture here are some comments referring to different concepts. We have already learnt that the concept "acceleration of a freely-falling body at 9.810 metres per second" means that the velocity of the body with respect to the centre of the earth is 9.810 metres greater when the earth has performed an additional 86400th part of its rotation. P.V.232-233 (P.L.204). The concept of sodium, which has also already been mentioned, can only be definitely applied to a body if it is soft as wax, easily cut, has a silver sheen on the cut surface, tarnishes easily, floats and thereby rapidly reacts with water, if it has the specific gravity 0.972, if it burns on ignition with a yellow flame and has an atomic weight of 23. Similarly, an animal will be subsumed under the concept 'whale' if it has the external form of a fish but turns out, after thorough anatomical investigation, to have double circulation, pulmonary respiration and all the other characteristics of the class of mammals. Again, the physicist subsumes under the concept "electro-magnetic current of unit intensity" that galvanic current which, acting with a magnetic horizontal component of $H = 0.2$ on a magnetic needle suspended in the centre of a circular wire of radius 31.41 cm,

through which the current has been made to pass, turns that needle 45 degrees out of the meridian. This presupposes a further set of operations for determining H. (W.L.417/418). And because 'oxygen' too is a concept which is not exhausted by an intuitive presentation but only by a definition which contains and sums up a number of experiences, and because this is true of all other physical concepts we can say that a concept is no more than a rule for producing an idea of certain properties. In the definition of a concept only those reactions are taken into account which are sufficient to determine the concept; other reactions of which it is already known that they are inevitably tied to those in the definition need not be specifically mentioned. E.u.I.127 (K.& E.94) What is then characteristic of concepts is the result of the reactions – manual or intellectual, analytic or constructive – prescribed by the definitions. Thus a body is electric if it exhibits certain sensory characteristics in certain reactions. A copper body is one whose bluish-green solution in dilute sulphuric acid exhibits a certain sort of behaviour when suitably treated. Etc. W.L.419-420. These reactions and the often very complicated activities required to produce them can only appear gradually and one after the other. 'Whether a mechanical system represents equilibrium or motion can be decided only through complex activities . . . But if one is aware that one can carry out this test at any time then one knows that a case of equilibrium will yield zero or a negative sum while a case of motion yields a negative sum and possesses the concept of work and can thereby distinguish positive from negative cases. Every physical and chemical concept may be expounded in this way. The object corresponds to the concept if it yields the expected reaction when tested: according to the circumstances this may be merely a matter of looking, or a complicated mental or technical operation and the ensuing reaction may be a simple sensation or a complicated process." E.u.I.131/132 (K.& E.97).

41 E.u.I.445 (K.& E.353/354).

42 As is the case in Hertz's mechanics.

43 M.282f. (M.320f.).

44 E.d.A.35; M.547.

45 M.237 (M.272-273).

46 A.d.E.267f. (A.S.343f.).

47 A.d.E.273 (A.S.350). Cf. P.V.233, E.u.I.426f., E.d.A.57.

48 M.547 (M.604).

49 E.d.A.57.

50 M.238 (M.274).

51 M.244 (M.281).

52 M.247.Cf.M.249 (M. 286,cf.288).

53 A.d.E.258 (A.S.331).

54 e.g.E.u.I.282 (K.& E.212).

55 The term *Moment*, or 'moment', is employed by philosophers influenced by Brentano such as Stumpf, Meinong and Husserl in a sense distinct from that it has in physics. It refers to individual, as opposed to general, properties and relations.–*Tr.*

56 M.548 (M.605).

57 A.d.E.275 (A.S.351).

58 Cf. also P.V.234 (P.L.205).

59 Cf. E.u.I.435 (K.& E.347); M.524.

60 A.d.E.(A.S.) Preface.

V. The final component of the concept of 'functional connexion' completed: the denial of natural necessity. The theory of elements. Final contradictions.

1 W.L.432ff.
2 W.L.457.
3 A.d.E.72 (A.S.86/87).
4 W.L.437.
5 Cf.: "What we call effect and cause are salient features of an experience which are important for our reproduction of the same in thought. Their importance wanes and the attention is transferred to new features the moment the experience in question becomes familiar. If the connexion of such features strikes us as a necessary one, it is simply because the interpolation of certain intermediate links with which we are very familiar and which therefore possess greater authority for us has often been sucessful in the past." P.V.227/228 (P.L.198/199).

"Once a fact becomes familiar we no longer require that its connecting marks be put into relief. Our attention is no longer attracted to the new and surprising and we cease to speak of cause and effect. Heat is the cause of the tension of steam; but when the phenomenon becomes familiar we think of the steam together with the tension proper to its temperature. Acid is the cause of the reddening of tincture of litmus; but later we think of the reddening as a property of the acid." M.524 (M.580).
6 E.d.A.31/32 (CE 55/56).
7 M.367 (M.415)
8 P.V.223 (P.L.194). Similarly M.6 (M.6).: "In the infinite variety of nature many ordinary events occur; while others appear uncommon, perplexing, astonishing or even contradictory to the ordinary run of things. . . When once one has reached the point where one is everywhere able to detect the same few simple elements, combining in the ordinary manner, then they appear to us as things that are familiar, we are no longer surprised, there is nothing new or strange to us in the phenomena, we feel at home with them, they no longer perplex us, they are explained."
9 M.16 (M.19).
10 P.V.239/240 (P.L.210/211).
11 M.16 (M.19).
12 M.12 (M.15).
13 M.77 (M.86).
14 M.75 (M.86).
15 E.u.I.312 (K.& E.234).
16 A.d.E.263 (A.S.337). Cf.W.L.435.
17 M.80/82 (M.91/94).
18 "A proposition in natural science" – we read elsewhere – "like any proposition in geometry, is of the form 'if M exists then N exists', where M and N can be more or less complicated groups of characteristics of phenomena and one group determines the other. Such a proposition may result directly from observation or indirectly through reflection and mental comparison of already known observations. . . The proposition 'if M exists then N exists' may be derived or explained from propositions expressing facts already known by means of a series of intermediate propositions. . . Thus Galileo explains the floating of very heavy dust in water and air in terms of the low speed of fall because of the large resistance produced by the fine distribution, Huygens completely derives the motion of pendulums from Galileo's mechanical principles", etc.E.u.I.262 (K.& E.195). Another example: "Arago found that a rotating copper disc (A) moves a magnetic needle (B) with it. As a result of Faraday's later discovery new elements (C) are interpolated between (A) and (B).

The discovery was that, in parts of the conductor which are moved relatively to the magnet, currents are generated which (according to Oerstedt) exert forces upon the magnet, and these forces (according to Lenz) act in the opposite direction to the motion. The connexion of A and B is elucidated by C, which, however, involves constructions of just the same kind. If C had been known previously, not only partially but wholly, deduction would have led to the discovery of the connexion of A and B". W.L.450. From this, moreover, it is concluded that the 'process of discovery' by deduction differs in no essential way from that by induction. W.L.449.

19 W.L.437. Elsewhere Mach strongly emphasises that an explanation can only be regarded as sussessful and a problem only ceases to exist when the right sides of the fact are taken into account, those which provide a simple unified conception. Thus: "We soon become familiar with the motion of isolated heavy bodies, but if a lighter body is raised by a heavier one, as on a pulley, we learn to notice the relations between several bodies and their weights. If we add findings from levers with unequal arms or from other machines, we are driven to consider not only the weights but their corresponding displacements in the direction of gravity and to see the products of their measures, that is, work done . . . Heavy bodies, when projected, may rise or fall, two cases that Aristotelian physics treated as different. Galileo takes notice of the acceleration of the motion, which makes all these cases similar and equally intelligible." E.u.I.264/265 (K.& E.196/197). Or, "For example, we see, quite contrary to the common run of our experience, a lever or pulley lifting a large weight by means of a small one. We seek the differentiating moment which the fact itself as given to the senses does not disclose to us. It is only when, comparing various similar facts, we have noted the influence of the weights, and of the arms of the lever, and by our own exertions have acquired the abstract concepts of 'moment' or 'work' that the problem is solved. 'Moment' or 'work' is the differentiating element. When it has become a habit of thought to pay attention to 'moment' or 'work' the problem no longer exists. "A.d.E.249 (A.S.320). And elsewhere: "We see a body thrown upwards. It rises. Why does it not now seek its place? Why does the velocity of its 'forced' motion decrease while that of the 'natural' motion of falling increases? Galileo, following up both facts, saw in both cases the same increase of velocity towards the earth. With this perception the problem was solved. Thus, not a place, but an acceleration towards the earth is assigned to bodies . . . Adhering to this new habit of thought, Newton saw the moon and planets moving like projected bodies but yet with peculiarities, which compelled him again to modify a little this habit of thought. Bodies, or rather their parts, do not maintain a constant acceleration towards one another. But they 'attract' each other with forces varying inversely as the square of their distances from one another and directly as their masses. This idea, which includes that of terrestrial heavy bodies as a special case, is very different from the original one . . . This process of transformation consists of two parts. On the one hand it consists in finding new identical characteristics in apparently different facts. On the other hand it consists in noting distinguishing characteristics in facts which have not hitherto been held to differ. In this way it becomes possible on the one hand to comprehend a constantly enlarging domain of facts with the same kind of habit of thought; and, on the other hand, to make variation of the habit of thought correspond to distinctions amongst the facts. This development is only a special case of a universally distributed biological process." W.L.385f.

20 E.d.A.31.

21 P.V.282/283 (P.L.255/256).

22 W.L.121.

23 W.L.458 f.

24 M.526 (M.582).

25 W.L.454/455. Cf. A.d.E.261. M.533. W.L.119,363.

26 Cf. p. 29–30 of the present work.

27 On laws: "One often speaks of laws of nature. What does the expression mean? The usual

opinion will be that the laws of nature are rules, which processes in nature must obey, resembling civil laws. A difference is usually seen in that civil laws can be broken while deviations from the laws of nature by natural processes are regarded as impossible. However, this view of the laws of nature is shaken by the reflection that we read off and abstract these laws from those processes themselves and that in doing this we are by no means immune to error." E.u.I.441 (K.& E.351).

28 M.280(M.318).Cf. also E.u.I.140 (K.& E.102): "Logical deductions from concepts remain intact so long as we retain those concepts; but the concepts themselves must always expect correction by the facts."

29 A.d.E.73 (A.S.89).

30 cf. Mach's own words: "The business of physical science is the reconstruction of facts in thought, or the abstract quantitative expression of facts. The rules which we form for these reconstructions are the laws of nature. In the conviction that such rules are possible lies the law of causality". M.547 (M.604).

31 M.547 (M.604).

32 M.549 (M.606).

33 M.523 (M.579).

34 M.79 (M.90) "As a rule it is a particular side or property of a fact which is of practical interest. Investigation is confined to this property. Facts which agree in possessing this property are treated as the same or of the same sort; those which differ in this property are treated as dissimilar . . . Practical needs impel us to abstraction." W.L.452. The decisive role of abstraction in enquiry is obvious. We can neither keep track of all the details of a phenomenon nor would it be sensible to do so. We take notice of those features that are of interest to us, and of those that appear to depend on these. The enquirer's first task is thus to compare different cases and set aside as incidental or irrelevant for the purpose in hand everything which has no bearing on his enquiry. This process of abstraction does in fact lead to very important discoveries." E.u.I.135 (K.& E.99/100)

35 "In the economical schematisation of science", then, "lies both its strength and its weakness. Facts are always represented at a sacrifice of completeness and never with greater precision than fits the needs of the moment." P.V.235 (P.L.206).

36 E.u.I.447 (K.& E.355/356).

37 E.u.I.189 (K.& E.140). Ibidem: "An important process consists in mentally diminishing to zero one or several conditions that quantitatively affect the result, so that the remaining factors alone must be taken as having influence. Physically, such a process is often impossible to carry out, so that we may speak of it as an idealization or abstraction. By considering the resistance to motion of a body impelled on a horizontal plane or the retardation of a body moving up a very slightly inclined plane as the angle becomes vanishingly small, we reach the idea of a body moving uniformly without resistance. In practice this case cannot be realized." Cf. M.306 (M.347) "Rest is only a very infrequent and indeed never completely realised case of motion . . . When, however, we occupy ourselves with cases of equilibrium, we are concerned simply with a schematic reproduction in thought of the mechanical facts. We then deliberately neglect these disturbances, displacements, bendings and tremors as without any interest for us."

W.L.454/455: "A further means [for the familiar application of scientific constructions] consists in the simplification and schematization of facts i.e., in their representation in pictures which contain the essential features and without anything superfluous that might distract the attention. Thus we think of a planet as a point and the path of an electric current as a line." Cf. also E.u.I.137 (K.& E.101),384.

38 M.133 (M.161).

39 P.V.228, cf. A.d.E.262.

40 E.u.I.449 (K.& E.357).

41 E.u.I.402 (K.& E.316), cf. W.L.456.

42 M.33 (M.40) The comparison with geometry is the source of characteristic statements elsewhere. Cf. E.u.I.376 (K.& E.290): "Physical metrical experiences, like all experiences forming the basis of experimental sciences, are idealized in concepts. The need to represent the facts by simple perspicuous concepts under easy logical control, is the reason for this. Absolutely rigid, spatially invariable bodies, perfect straight lines and planes no more exist than a perfect gas or a perfect liquid. Nevertheless, we work by preference and more readily with these concepts than with others that conform more closely to the properties of the objects, deferring the consideration of the deviations. Theoretical geometry does not even need to consider these deviations inasmuch as it assumes objects that fulfil the requirements of the theory absolutely, just as theoretical physics does." E.u.I.407 "Like physics, geometrical theory is simpler and more exact than can be guaranteed by experience and its accidental disturbances."

43 E.u.I.140 (K.& E.102).

44 Cf. A.d.E.260.

45 E.u.I.447/448 (K.& E.356) Cf. E.u.I.302/303.

46 Cf. p. 71 of the present work.

47 W.L..461/462.

48 Of the considerations which lead Mach to the theory of elements we want, however, to note two: on the one hand he relies on the difficulties which result from the dualist conception, in particular the absence of any prospect of explaining sensations by reference to the motions of atoms in a nervous substance; on the other hand, the tendency which is actually present in science towards unification of individual disciplines. One need only think of optics, the theory of electricity and magnetism, physical chemistry, physiological chemistry etc.; the search for ideas which integrate and embrace these different domains is a natural further step. "It may be", he says, "that the physicist is still satisfied with the notion of a rigid matter . . . the physiologist or psychologist can make nothing of this at all. But any one who has in mind the integration of the sciences into a single whole has to look for a conception to which he can adhere in every scientific domain. Now if we resolve the whole material world into elements which at the same time are also elements of the psychical world and, as such, are commonly called sensations; and if we regard it as the sole task of science to inquire into the connexion and combination of these elements, which are of the same nature in whatever scientific domain, and into their mutual dependence on one another – then we may justifiably expect to build a unified monistic structure upon this conception and thus get rid of the distressing confusions of dualism. Indeed it is by regarding matter as something absolutely stable and immutable that we actually destroy the connexion between physics and physiology . . . But when it is a question of bringing into connexion two adjacent domains, each of which has been developed in its own particular way, the connexion cannot be effected by means of the limited conceptions of a narrow special domain. Here more general considerations must lead to the creation of more general concepts which are adequate for the wider domain." A.d.E.242/243 (A.S.312/313). On the two considerations, cf. inter alia A.d.E.1(1),23-26 (29-32) 36 (44),37 (45),46 (56),188 (243),257 (329), 258 (330),283 (365-366); E.u.I.3 (2-3), 234, 451; P.V.237, 241, 285; M. 504 (M.506). Both groups of objections touch on difficulties which, deserving of attention though they are, must of course be disregarded here as inconclusive in view of the numerous different and incomplete attempts to solve this problem.

49 W.L.396. Cf. A.d.E.245 (A.S.314-315), 246 (A.S.316).

50 A.d.E.255 (A.S.316), 268 (A.S.328).

51 A.d.E.267/268 (A.S.343). Cf.P.V.235.: In reality it is always a matter of deriving one part of a phenomenon from another. In this process our ideas must be directly based on sensations. This we call measurement.

52 M.547 (M.502) P.V.234.

53 A.d.E.246 (A.S.342).

54 M.547 (M.502).

55 W.L.404.Cf.inter alia: P.V.220 (P.L.191),235 (P.L.206),236 (P.L.207); M.133,505,504, E.u.I.126,139,311 (K.& E.92,102).

56 "Such a concept word in science has the purpose of reminding us of the combination of all the object's reactions as denoted in the definition, in order to draw these memories into consciousness as though by a thread . . . Of course every definition may contain further concepts, so that only the last and ultimate conceptual building bricks can be resolved into the palpable reactions which are their marks or features." E.u.I.127, (K.& E.93).

57 A.d.E.35-37 (A.S.41/42).

58 P.V.240/241, (P.L.211/212); A.d.E.23f. (A.S.29f.).

59 E.u.I.14 (K.& E.9).

60 M.504/505 (M.559).

61 A.d.E.1/2 (A.S.2).

62 M.523 (M.579).

63 P.V.229 (P.L.200).

64 A.d.E.2 (A.S.2).

65 A.d.E.9/10 (A.S.12).

66 A.d.E.5 (A.S.6).

67 A.d.E.256 (A.S.329).

68 A.d.E.256 (A.S.329).

69 P.V.229 (P.L.200).

70 A.d.E.10 (A.S.12).

71 M.523 (M.579); cf. A.d.E.23 (A.S.29).

72 A.d.E.23ff. (A.S.29ff.).

73 A.d.E.258 (A.S.331).

74 A.d.E.1/2 (A.S.2).

75 Mach assumes that feelings and the will are composed of traces of sensations – A.d.E.11,17,82; E.u.I.9 – and that sensations are diffusely localised sensations – E.u.I.18f. (K.& E.17f.). "But", he says, "should it prove impossible to manage with only one sort of such elements then more will have to be assumed". A.d.E.17 (A.S.22).

76 A.d.E.8 (A.S.9).

77 Examples of physical connexions: a white ball falls upon a bell, there is a sound, the ball turns yellow before a sodium lamp, red before a lithium lamp. Psychological connexions: the ball turns yellow on our taking santonine, disappears if we close our eyes and becomes two balls if we press one eye to the side. A colour, then, is a physical object if we consider its dependence on other colours, temperatures, spaces etc. and a sensation if considered as dependent on the retina. All that has changed is the direction of the investigation. Whether elements are physical objects or sensations depends on the relations of functional dependence they enter into. A.d.E.11/14 (A.S.13-18). It is only in virtue of a different sort of connexion that $\alpha\,\beta$ . . . appear in a different domain than A B . . .; "that the elements α or A . . . appear in a different domain means, if we go to the bottom of the matter, simply this, that these elements are united with other different elements." This is said to be the origin of the difference between perception and representation of the same object. A.d.E.16 (A.S..20/21); cf. A.d.E.26ff.,35f.,42-46,241, (A.S.33ff.,42f.,50-56).

78 What is true of bodies is true also of the ego. It is not the ego which is primary but the elements of which it is composed; the ego does, however, then react to these sensations. That I sense green means only that the element green occurs in a certain complex of other elements (sensations, memories). "When I cease to sense green, when I die, then the elements no longer occur in the familiar combinations. That is all. Only an ideal unity pertaining to the economy of thought has ceased to exist, not any real unity." A.d.E.19,21.

79 E.u.I.9 (K.& E.6).

80 A.d.E.29 (A.S.36) It should be noted that this investigation of functional connexions is,

according to Mach, the only possible way of fathoming reality. A.d.E.23ff.,29/30; E.u.I. 10/11.

81 A.d.E.11-14 (A.S.13-18): A powerful analogy as well as practical needs force us, according to Mach, to think of memories, fears and wills as connected with the bodies of other people and animals. It is the behaviour of other people which forces me to assume that my body and other bodies are immediately present to them and that my memories and wishes are the object of an irresistible analogical inference for them. When we investigate the influence of our bodies on our sensations we complete observed facts by analogy; the same is true when we infer the sensations of others on the basis of observation of their behaviour. In the first case the completion is physical (involving the nerves) and so more familiar, but there is no essential difference.

At E.u.I.6 (K.& E.5) The totality of what is immediately given in space for everyone is called the physical and what is given only to one, while others must infer it by analogy, is called the psychical.

A.d.E.26ff. (A.S.33ff.) The elements A B C . . . appear directly as external to the elements K L M . . . The world of things appears to be independent of the ego because the dependence of A B C . . . on the complex K L M . . ., which is continually repeating itself in the same way, is disregarded. The thought of the ego is formed by paying attention to the properties of K L M . . . and its connexion with α β γ . . . Further, other bodies K′ L′ M′, K″, L″. M″ etc. behave in such a way that their behaviour in contrast to that of A B C . . ., only becomes familiar if $\alpha'\beta'\gamma'$ . . . like $\alpha''\beta''\gamma''$ . . . are thought of as connected with them. The sensations of other people are completed on the basis of analogy but to ascribe to sensations a nature different from that of K L M . . . A B C . . . on this account is unnecessary and misleading.

A.d.E.35 (A.S.42): It is possible to follow the course of a physical (physiological) process through a nerve. But the behaviour of the organism in question can be predicted with much greater accuracy i. e. one understands it better, if sensations and memories etc. are attributed to it. In doing this one completes what is observed by something which is not to be met with in the realm of one's own sensations. This antithesis is not as absolute as it appears. For, first, the physicist frequently completes complexes of sensations by elements which at the time are not observed and he does so on the basis of analogy. He takes the moon, for example, to be tangible, heavy, slow and so does every day precisely what here seems so strange. Secondly, the abruptness of the opposition disappears once the following observation is borne in mind. Consider the leaf of a plant. Its greenness (A) is connected with an optical sensation of space (B), with a sensation of touch (C) and with the visibility of a source of colourless light (D). If the yellow (E) of a sodium flame takes the place of (D) the green (A) will give way to brown (F). If the leaf is treated with alcohol – "an operation which, like the preceding one, can be represented by sensory elements" – the green (A) will give way to white (G). These are all physical observations. But (A) is also connected with a certain process of my retina. This too I can reduce to elements X Y Z . . . by investigating someone else's eye and transferring the results on the basis of analogy. Now in its dependence on B C D E. . . A is a physical element, in its dependence on X Y Z . . . a sensation and can also be considered as a psychical element. "The green (A), however, is not altered at all in itself whether we direct our attention to the one or to the other form of dependence."

82 A.d.E.22 (A.S.27).

83 A.d.E.26f. (A.S.33f.).

84 A.d.E.29 (A.S.36).

85 A.d.E.269 (A.S.344).

86 A.d.E.27 (A.S.33/34).

87 [Cf. C. Stumpf, *Erscheinungen und psychische Funktionen,* 1906, Abh. d. König. preuss. Ak. d. Wiss., phil.-hist. Kl., 4 – *Tr.*]

[88] E.u.I.277 (K.& E.207).
[89] E.u.I.28 (K.& E.22).
[90] M.6 (M.5).
[91] E.u.I.450/451 (K.& E.358).
[92] P.V.250 (P.L.221).
[93] W.L.393.
[94] E.u.I.446,449/450 (K.& E.354,357).
[95] W.L.454.
[96] E.u.I.227/228 (K.& E.208).
[97] M.293 (M.331).
[98] M.280 (M.318).

Analytica

The series *Analytica* serves the publication of historical and systematic investigations in logic, ontology and the philosophy of language. Two principal currents can be discerned in the history of these philosophical disciplines. The first was initiated by Aristotle and continued into the modern period by his commentators, by Boethius, and by the first and second scholastics. After an interruption brought about by Kant and the German idealists, the tradition was reestablished in the 19th century, being represented above all by Brentano and his students, Husserl, Meinong and Twardowski.

The second, mathematical or formal current – foreshadowed in the work of Leibniz – was initiated during the second half of the 19th century; by Boole, Schröder and Frege in the field of logic, and by Frege again, and Cantor, the founder of set theory, in ontology. In the field of the philosophy of language the rational grammars developed in the pre-mathematical period have given way, in our own century, to mathematical linguistics and philosophical logic.

The traditions of philosophy in the spirit of Aristotle and of mathematical logic have too often been held to be entirely alien to each other. It is one principal aim of the present series to encourage the collaboration of the scholars of these two traditions.

Hans Burkhardt
Logik und Semiotik in der Philosophie von Leibniz
488 pp., index, cloth
ISBN 3-88405-001-X

M. Richard Diaz
Topics in the Logic of Relevance
144 pp., index, cloth
ISBN 3-88405-003-6

Larry Hickman
Modern Theories of Higher Level Predicates
Second Intentions in the Neuzeit
191 pp., index, cloth
ISBN 3-88405-000-1

Herbert Hochberg
Logic, Ontology and Language
Essays on Truth and Reality
appr. 544 pp., index, cloth
ISBN 3-88405-030-3

Barry Smith, Editor
Parts and Moments
Studies in Logic and Formal Ontology with contributions by W. Künne, K. Mulligan, G. Null, P. Simons, R. Simons, B. Smith and D. Willard
564 pp., index, cloth
ISBN 3-88405-012-5

Peter Rutz
Zweiwertige und mehrwertige Logik
Ein Beitrag zur Geschichte und Einheit der Logik
108 pp., index, cloth
ISBN 3-88405-039-7

Paul Thom
The Syllogism
312 pp., index, cloth
ISBN 3-88405-002-8

Available titles December 1982
In North America distributed by
John Benjamins North America, Inc.
One Buttonwood Square, 202
Philadelphia, PA 19130/USA

Philosophia Verlag · München · Wien

Philosophia Resources Library

With the growing interest in Austrian philosophical thought both in the German speaking world and in England and America there has arisen the need to make once more accessible the classical texts of this unique intellectual tradition.
Philosophia Resources Library offers new editions of principal works in the Austrian tradition, with introductions and commentaries by leading specialists in the field. It will include English translations of selected items and will be complemented by monographs and studies on specific themes.

Austrian Philosophy
Studies and Texts
edited by J. C. Nyíri with contributions by L. Congdon, A. Janik, W. Johnston, R. Haller, F. Rossi-Landi, J. Eötvös and R. Musil
200 pp., cloth
ISBN 3-88405-004-4

Christian von Ehrenfels
Philosophische Schriften · Band I
Werttheorie
edited by R. Fabian
introduction by W. Grassl
615 pp., index, cloth
ISBN 3-88405-033-8

Christian von Ehrenfels
Philosophische Schriften · Band II
Ästhetik
edited by R. Fabian
introduction by R. Haller
appr. 512 pp., index, cloth
ISBN 3-88405-032-X

Roman Ingarden
Man and Value
translation from the Polish and German by A. M. Szylewicz
appr. 256 pp., index, cloth
ISBN 3-88405-042-7

Roman Ingarden
Selected Papers in Aesthetics
edited and with an introduction by Peter McCormick
appr. 288 pp., index, cloth
ISBN 3-88405-042-7

Thomas G. Masaryk
Der Selbstmord als sociale Massenerscheinung der modernen Civilisation
Reprint of the edition Vienna: 1881 with introduction in English and German by J. C. Nyíri
42 + XVI + 245 pp., cloth
ISBN 3-88405-014-1

Kasimir Twardowski
Zur Lehre vom Inhalt und Gegenstand der Vorstellungen
Eine psychologische Untersuchung
Reprint of the edition Vienna: 1894
introduction by R. Haller
142 pp., index, cloth
ISBN 3-88405-017-6

Titles available or in preperation
December 1982

Philosophia Verlag · München · Wien

The International Carl Menger Library

This new series has been inspired by the current renaissance of the Austrian School of Economics and the revival of interest in its philosophical and historical background. Through it, formerly inaccessible works of a long neglected school of thought will be made available to a wider public in both English and German.
Reprints and translations of important older works, as well as original writings by authors working in the Austrian tradition, will clarify and elaborate this new and dynamic alternative to the conventional wisdom in economics.

Friedrich A. von Hayek
Geldtheorie und Konjunkturtheorie
reprint of the edition Vienna: 1929
XVI + 160 pp., paper
ISBN 3-88405-008-7

Friedrich A. von Hayek
Individualismus und wirtschaftliche Ordnung
reprint of the first edition Erlenbach-Zürich: 1952
bibliography by K. R. Leube
358 pp., paper
ISBN 3-88405-007-9

Friedrich A. von Hayek
Mißbrauch und Verfall der Vernunft
reprint of the edition Frankfurt: 1959
VI + 360 pp., paper
ISBN 3-88405-009-5

Margit von Mises
Ludwig von Mises
Der Mensch und sein Werk
with 42 fotographs on plates
312 pp., index, cloth
ISBN 3-88405-024-9

Available titles December 1982

Ludwig von Mises
Nationalökonomie
Theorie des Handelns und Wirtschaftens
reprint of the edition Geneva: 1940
XVI + 756 pp., index
cloth edition ISBN 3-88405-010-9
paper edition ISBN 3-88405-023-0

Ludwig von Mises
Die Gemeinwirtschaft
Untersuchungen über den Sozialismus
preface by Chr. Watrin
reprint of the edition Jena: 1932
8 + xx + 500 pp., index
cloth edition ISBN 3-88405-018-4
paper edition ISBN 3-88405-025-7

Leo Schönfeld-Illy
Grenznutzen und Wirtschaftsrechnung
reprint of the edition Vienna: 1924
preface by Kurt R. Leube
VII + 209 pp., cloth
ISBN 3-88405-040-0

Richard von Strigl
Kapital und Produktion
reprint of the edition Vienna: 1934
preface by Ludwig M. Lachmann
updated bibliographic references by Ulrich Fehl
11 + X + 247 + 5 pp., cloth
ISBN 3-88405-037-0

Philosophia Verlag · München · Wien